Biomedical Applications of Mesoporous Ceramics

Drug Delivery, Smart Materials and Bone Tissue Engineering

T0186347

Biomedical Applications of Mesoporous Ceramics

Drug Delivery, Smart Materials and Bone Tissue Engineering

María Vallet-Regí

Miguel Manzano García

Montserrat Colilla

CRC Press
Taylor & Francis Group
Boca Raton London New York

CRC Press is an imprint of the
Taylor & Francis Group, an **informa** business

CRC Press
Taylor & Francis Group
6000 Broken Sound Parkway NW, Suite 300
Boca Raton, FL 33487-2742

First issued in paperback 2019

© 2014 by Taylor & Francis Group, LLC
CRC Press is an imprint of Taylor & Francis Group, an Informa business

No claim to original U.S. Government works

ISBN-13: 978-1-4398-8307-5 (hbk)
ISBN-13: 978-0-367-38060-1 (pbk)

Library of Congress Cataloging-in-Publication Data

Vallet-Regí, Maria.
  Biomedical applications of mesoporous ceramics : drug delivery, smart materials, and bone tissue engineering / Maria Vallet-Regi, Miguel Manzano Garcia, Montserrat Colilla.
      p. ; cm.
  Includes bibliographical references and index.
  ISBN 978-1-4398-8307-5 (hardback : alk. paper)
  I. Manzano Garcia, Miguel. II. Colilla, Montserrat. III. Title.
  [DNLM: 1. Ceramics--therapeutic use. 2. Bone Substitutes. 3. Drug Delivery Systems. 4. Silicates--therapeutic use. QT 37.5.C4]

  610.28'4--dc23   2012031219

Visit the Taylor & Francis Web site at
http://www.taylorandfrancis.com

and the CRC Press Web site at
http://www.crcpress.com

Contents

Preface

The advances in biomedical applications of mesoporous ceramics have been very important in the last decade. A great part of this relevance has been due to the search for multifunctional roles for these well-known materials. To achieve that multifunctionality with already known ceramics, it has been necessary to design new synthetic and/or conformation processes.

The present book is focused on mesoporous ceramics for biomedical applications, so ceramics with porosity ranging from 2 to 50 nm will be deeply studied. The main biomedical application of these mesoporous ceramics is in the field of drug delivery, which does not impede their use as starting material for the fabrication of scaffolds for tissue engineering and, therefore, performing two roles simultaneously.

All ceramics, commonly dense, can be produced with a mesoporous structure following the appropriate synthetic method, which is based on the formation of templates with micelles on which ceramics can be precipitated. For this reason, Chapter 1 of this book will describe biocompatible and bioactive ceramics, which can be obtained with a mesoporous structure following the adequate synthetic path. As an example of this, ordered mesoporous glasses will be fully described, from their synthesis to their applications in biomedicine.

Supramolecular chemistry using surfactants has facilitated the possibility of inducing mesopores to a great variety of bioceramics. Among them, this book will describe templated glasses, also known as ordered mesoporous glasses, hydroxyapatites, and mesoporous silica, but it is also possible to produce many other materials following this route.

The chemistry in confined spaces, which was performed in the 1970s with all the zeolites technology, was the root of a great amount of experimental work. In 1972, Mobil Corporation developed a procedure to transform methanol into gasoline using ZSM-5 (Zeolite Socony Mobil) as catalyst.

Inspired by this innovative process, many zeolites with larger pore size were targeted to introduce larger and cheaper alcohol molecules. The aim was to produce cheaper gasoline through acid-base reactions produced in the small cavities of those zeolites. Although it did not work using zeolites, after further research on materials with large pores, the first mesoporous silica material was produced. The synthesis by Mobil in the United States and the research group headed by Kuroda in Japan occurred at the same time. Those mesoporous silica materials with large pores were obtained using the supramolecular chemistry of surfactants, as will be detailed in Chapter 1.

Although mesoporous silica materials were initially designed for catalysis applications, their utility for other applications in diverse research areas such as magnetism, sensors, optical materials, photo catalysis, fuel cells, thermo electrics, and even in health research, was soon recognized.

In 2001, these materials were proposed as drug delivery systems for the first time. The very same mesoporous ceramic matrices that were able to transform alcohol

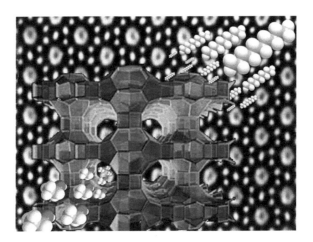

FIGURE 1   (See color insert.) Schematic representation of the ZSM-5 catalyst, which is made of silica and alumina and was employed to transform alcohol into gasoline.

into gasoline were employed to adsorb and release pharmaceutical agents. The lack of toxicity is a mandatory condition to employing these materials as drug delivery systems, which will be discussed in Chapters 2, 4, and 5.

One of the great advantages of these ordered mesoporous materials is the great versatility of the synthetic process, which allows the production in bulk as well as in

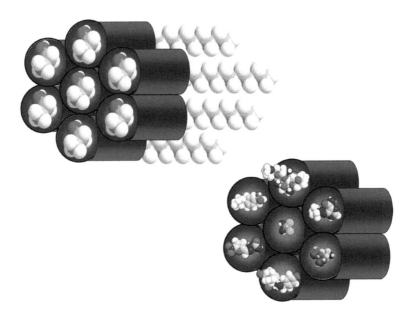

FIGURE 2   (See color insert.) Schematic representations of the conversion of alcohol into gasoline (top) and confinement of different pharmaceutical agents into the mesopores of the silica matrices.

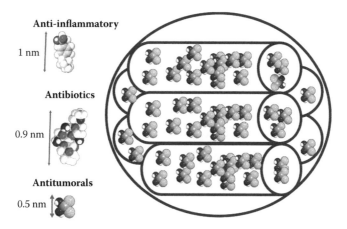

FIGURE 3 Schematic representation of confinement of different drugs, such as diverse anti-inflammatory, antibiotics, or antitumor therapies into the mesopores of the silica nanoparticles.

microcapsules and even as nanoparticles, expanding their application areas, as will be detailed in Chapters 3 and 4.

The research on bioceramics started back in the 1950s, when the research was focused on inert ceramics so when in contact with living tissues there would be no reaction, thereby minimizing the adverse effects. Although biocompatibility of those materials was also aimed for, their interaction with the biological environment where they would be placed was not taken into account. The research was more focused on finding better mechanical properties for those ceramics, which are well known to be very brittle. The trend to achieving improved mechanical properties was to compact and densify those ceramics as much as possible. However, those ceramics always presented a foreign-body response once they were implanted, being surrounded by an acellular collagen layer, called a fibrous capsule, acting as an interface between the ceramic implant and the living tissues.

In the 1980s, other alternatives within the bioceramics research area were explored. It was found that calcium phosphates, glasses, and glass-ceramics were able to bond to osseous tissue without the formation of the fibrous capsule mentioned earlier. This capacity to directly join to living tissues is called bioactivity, and it brings out a totally new generation of bioceramics. However, the mechanical properties of this new generation of bioceramics were very poor, so despite their excellent bioactivity, these materials were produced as dense as possible. The consequence of this trend is that the excellent bioactive properties could be developed only at the surface of the implant, not in the inside part. The reason for this was that because of high density, it was not possible to interchange any fluid within the physiological environment, together with the absence of oxygenation, vascularization, and, of course, the entrance of any cell. In this period of time, by the end of the 1990s, the researchers working on biomaterials realized that they were working in the wrong direction, without taking into consideration the biological world from the beginning of the material design.

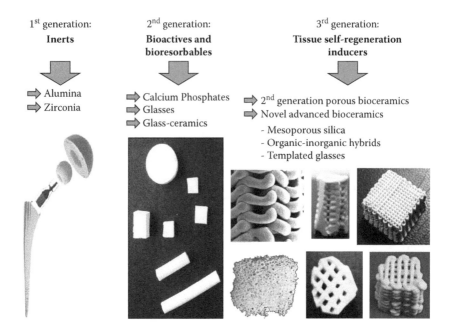

1st generation:

Inerts

⇨ Alumina
⇨ Zirconia

2nd generation:

**Bioactives and
bioresorbables**

⇨ Calcium Phosphates
⇨ Glasses
⇨ Glass-ceramics

3rd generation:

**Tissue self-regeneration
inducers**

⇨ 2nd generation porous bioceramics
⇨ Novel advanced bioceramics
   - Mesoporous silica
   - Organic-inorganic hybrids
   - Templated glasses

FIGURE 4   Scheme of the different generations of bioceramics together with some examples of those bioceramics employed in the clinic.

By the end of the 1990s, there was a new trend that consisted of designing ceramics with interconnected porosity and with a hierarchical porous structure to allow biological entities to develop their role into the ceramic core. These ceramics must present pores in very different ranges, from 1 µm up to more than 1000 µm. It is at this moment and under these requirements when new processing methods were implemented to design macroporous pieces with a previously designed porosity. These bioceramic pieces have to be composed of bioactive and/or degradable ceramics, so they can be resorbed at the same rate as the new bone tissue is formed. In this way, scaffolds for tissue engineering start to be produced. Consequently, stem and progenitor cells, biochemical agents that act as signals, growth factors, and scaffolds fabricated with natural and synthetic bioceramics, are the milestones for the third-generation biomaterials.

Third-generation bioceramics are designed to promote the self-regeneration of natural tissue and, at the same time, be an actual support of cells and inducting molecules. This generation of bioceramics includes those of the second generation, such as calcium phosphates, glasses, and glass-ceramics, but are processed with synthetic methods that make it possible for porous structures to develop certain biological functions. It is also the starting point in the design of new advanced bioceramics that follow those same requirements. Among them, this book will deal with organic-inorganic hybrids, mesoporous silica ceramics, and templated glasses. These mesoporous ceramics present additional porosity with a pore size ranging between 2 and 50 nm. This added value makes these materials very interesting for biomedical applications, since they can be designed to accomplish more than one function at a time, such as tissue engineering and drug delivery technologies.

Every bioceramic from the third generation employed in biomedicine should be biocompatible; biomimetic inspired; with a composition and structure similar to natural bone to achieve a perfect adaptability to the physiological context; bifunctional, so the implant will follow physical and mechanical requisites in the physiological environment; bioactive, so the implant will allow and promote the generation of new bone in the surrounding tissues, that is, have good osteoinduction and osteointegration; and finally, bioresorbable, so the implant will disappear from the biological environment through cellular activity and/or solubility in physiological fluids.

Consequently, bioactive ceramics that are osteoconductive, nontoxic, noninflammatory, and nonimmunological, and that can bond to osseous tissue can be applied in orthopedics and stomatology. Nowadays it is possible to produce constructs for tissue engineering, although this application is not commercially available yet. The traditional applications of these bioceramics are as coatings, small pieces, and filling material in both powder and bulk. The future applications of these bioceramics will be based on the knowledge of the osseous defect to then select the type and shape of the bioceramic to be processed, with the added value of combining micro- and mesoporosity. This possible combination opens the gates to a minimum of two simultaneous functions, bone repairing together with drug delivery with a local and controlled release in the implant site. For the former role, microporosity is necessary for bone tissue regeneration, while for the latter function, mesoporosity is a requirement to confine pharmaceutical agents into the mesopores.

Third-generation bioceramics induce the regeneration of tissues, stimulating the response of the cells responsible for the tissue self-regeneration. These bioceramics should have two components: a functional component to act as support, which should be porous and would be responsible for the activity and resorption in the ceramic; and an intelligent component, which would allow modification of its properties as a response to physicochemical variations of the surrounding environment, and would introduce biologically active substances, such as growth factors or hormones. This new generation of materials, which are porous, bioactive, and biodegradable, can be designed to stimulate specific cellular response at the molecular level to help to the body to self-repair. To achieve that, it is necessary to provide them with an added value so they can behave as smart materials. The way of doing this is through an organic functionalization to both promote the uptake of active molecules and interact with the surrounding biological tissues. When functionalizing the surface of these bioceramics, there is a challenge: obtaining surfaces that are able to entrap proteins without modifying the cells' activity.

The design of a nonimmunological scaffold similar to the extracellular matrix should be carried out to favor water retention, allow cellular retention on the scaffold, and provide adequate porosity to allow cells to grow in a 3D environment, to allow the flux of nutrients and oxygen along the growing cells. The research on bioceramics for cellular therapy is focused on designing methods to increase the efficacy through a strict control of the porous structure, so there would be an increment on cellular adhesion and protein adsorption. Additionally, the functionalization of the bioceramic surface with bioactive molecules such as enzymes, proteins, peptides, antibodies, etc., favors the migration, proliferation, and differentiation of the mesenchymal stem cells.

The present book is focused on mesoporous ceramics. With those materials, it is possible to design and fabricate ceramic implants for controlled release of drugs and substrates for tissue engineering. Both applications can be coordinated for a specific application, as will be detailed. Additionally, mesoporous nanoparticles have been proposed for many applications in nanomedicine, from cancer to gene therapy.

All these mesoporous ceramics with applications in the biomedical area are at the research stage, and in the near future they will be developed. The different chapters of this book will detail the research and development of these biomedical applications.

About the Authors

María Vallet-Regí was born in Las Palmas, Spain. She studied chemistry at the Universidad Complutense de Madrid (Spain) and received her PhD at the same university in 1974. She is a full professor of inorganic chemistry and head of the research group Smart Biomaterials in the Department of Inorganic and Bioinorganic Chemistry of the Faculty of Pharmacy at Universidad Complutense de Madrid. Dr. Vallet-Regí has written more than 600 articles and several books. She was the most cited Spanish scientist according to ISI Web of Knowledge, in the field of materials science in these past decades.

She is a Fellow of Biomaterials Science and Engineering at the International College of Fellows of Biomaterials Science and Engineering (ICF-BSE), Numbered Fellow of the Spanish Royal Academy of Engineering and the Royal National Academy of Pharmacy. She has received the Prix Franco-Espagnol 2000 from Societé Française de Chimie, the Spanish Royal Society of Chemistry (RSEQ) award in inorganic chemistry 2008, the Spanish National Research Award in engineering 2008, FEIQUE research award 2011 and the RSEQ research award and gold medal 2011.

Miguel Manzano was born in Madrid, Spain in 1976. He completed his studies of organic chemistry at Universidad Autónoma de Madrid, Spain in 2000 and received his PhD at the School of Biomedical and Biological Sciences from the University of Surrey, United Kingdom, in 2004. Returning to Spain, he obtained a position as assistant lecturer at the Department of Inorganic and Bioinorganic Chemistry of the Faculty of Pharmacy at Universidad Complutense de Madrid, Spain. In 2011, Dr. Manzano went to the Massachusetts Institute of Technology to complete a postdoctoral research fellowship for almost a year in the laboratory of Robert Langer at the Koch Institute, MIT. His research interests are focused on novel materials for biomedical applications with special interest in bone tissue engineering and drug delivery systems. He has had many articles and papers published in peer-reviewed international scientific journals.

**Montserrat Colilla** was born in Madrid, Spain, in 1975. She studied chemistry at Universidad Autónoma de Madrid and received her PhD degree at the same university in 2004, after completing a predoctoral fellowship at Instituto de Ciencia de Materiales de Madrid of the Spanish Council for Scientific Research (CSIC). In 2005, she moved to the Department of Inorganic and Bioinorganic Chemistry of the Faculty of Pharmacy at Universidad Complutense de Madrid, where she has held a position as associate professor since 2011. Her research interests are focused on bioceramics for bone tissue regeneration and drug delivery applications. She has many publications in international scientific journals and various book chapters in the field of organic-inorganic hybrid materials and ordered mesoporous materials for biomedical applications.

1 Biocompatible and Bioactive Mesoporous Ceramics

The term *biocompatibility* refers to the behavior of biomaterials in the living body. The most accepted definition of biocompatibility was given by D. F. Williams: "Biocompatibility refers to the ability of a biomaterial to perform its desired function with respect to a medical therapy, without eliciting any undesirable local or systemic effects in the recipient or beneficiary of that therapy, but generating the most appropriate beneficial cellular or tissue response in that specific situation, and optimizing the clinically relevant performance of that therapy" (Williams 2008, 2951). Attending to this definition, it is clear that all biomaterials, whether they will be used for drug delivery or tissue engineering, should present a biocompatible response.

Bioactivity is related to the type of interaction of a biomaterial to any cell tissue in the living body. In the context of bone tissue regeneration, bioactive materials are capable of bonding to living bone through the formation of bone-like apatite on their surface when implanted in the living body.

Ceramics materials are inorganic, nonmetallic solids that were traditionally prepared through a thermal treatment and a subsequent cooling. They have been employed in a wide variety of areas, such as construction and pottery, and even in more technological applications such as disk brakes and engine turbine blades. Bioceramics exhibit good biocompatibility, and they can be divided into two categories: *inert* (no reaction with living tissues), such as alumina or zirconia, and *bioresorbable-bioactive* (those that present bone bioactivity), such as calcium phosphates or glasses. Inert ceramics were developed in the 1960s for biomedical applications, while bioactive ceramics were mainly developed in the 1980s (Figure 1.1). Bioactive ceramics are commonly used as dental and bone implants because they bond readily to bone and other tissues in the body without rejection or inflammatory reactions.

In the 2000s, bioceramics were designed with porous structure bearing a microporosity equivalent to bone in an attempt to improve their efficiency for bone tissue regeneration, giving rise to mesoporous materials and organic-inorganic hybrids. The properties of those porous bioceramics depend on the porous size. The world authority on chemical nomenclature, the IUPAC, divides porous materials into three main groups: microporous (<2 nm), mesoporous (2–50 nm), and macroporous (>50 nm) (Sing et al. 1985). Thus mesoporous bioceramics are implantable materials that can be employed for controlled release of biologically active molecules, thanks to their mesostructure with pores ranging between 2 and 50 nm. These mesoporous

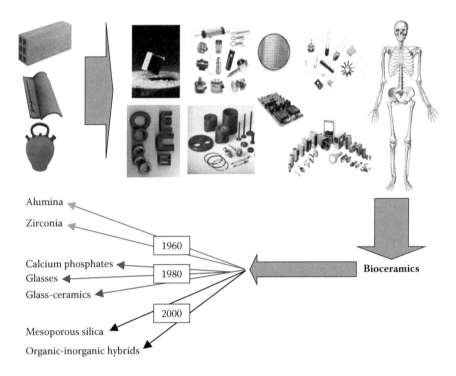

FIGURE 1.1   Traditional applications of ceramics (top) and chronological evolution of bioceramics (bottom).

bioceramics can be employed for the fabrication of scaffolds for bone regeneration, as schematically described in Figure 1.2. Mesoporous materials contribute to these scaffolds a dual benefit: regeneration thanks to their bioactivity, and biomolecule delivery thanks to their mesoporosity.

1.1   GENERAL PROPERTIES OF ORDERED MESOPOROUS CERAMICS

Ordered mesoporous ceramics are unique materials characterized by an ordered mesostructure of porous and disordered arrangement at the atomic level. They are produced using surfactants (amphiphilic organic molecules) as templates to direct the assembly and subsequent condensation of the inorganic precursors, leading to a network of cavities arranged periodically. The advances in the *sol-gel* chemistry, which is an inorganic polymerization process taking place in mild conditions that allows the association of mineral phases with organic or biological systems (Coradin, Boissiere, and Livage 2006), has fueled the development of these ordered mesoporous bioceramics. The reason is that the sol-gel process allows the induction of mesoporosity to traditional ceramic compositions.

Ordered mesoporous ceramics were reported for the first time in the 1990s, when the porous solids industry was trying to find materials with larger pores than zeolites

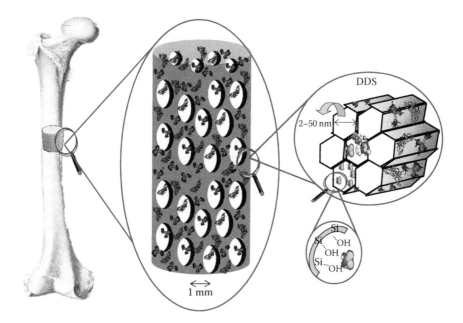

FIGURE 1.2 **(See color insert.)** Scheme of a potential scaffold made of mesoporous bioceramics and with bone regenerative character together with drug delivery capabilities.

(microporous class materials) to improve their applicability as adsorbents, catalysts, and catalyst supports. Japanese academic investigators (Yanagisawa et al. 1990) and Mobil Oil Corporation researchers (Kresge et al. 1992) started to employ surfactants as structure-directing agents to produce new types of material, KSW-n and M41S families of mesostructured materials, respectively.

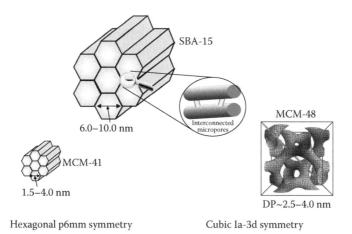

FIGURE 1.3 Schematic representation of some of the most popular ordered mesoporous materials and their respective mesopore diameters.

The porosity characteristics of these ordered mesoporous ceramics depend on the type of template employed during the so-called liquid-crystal templating (LTC) mechanism (Beck et al. 1992). This synthetic process is based on the dissolution of surfactant (surface active agent) molecules into polar solvents to form liquid crystals. When the amount of template dissolved in the solution is above the critical micellar concentration (cmc), the surfactant molecules will form aggregates called micelles. The characteristics of these micelles depend on the chemical nature of the surfactant employed and on the processing conditions, such as concentration, pH, temperature, etc. In the next step the micelles aggregate themselves to form supramicellar structures with specific geometries, which depend on the experimental conditions. These supramicellar geometries, such as hexagonal, cubic, laminar, etc., are the basics of the geometry of the resulting mesoporous framework. At this point of the process, the ceramic precursors are added to the liquid crystals, and the sol-gel process takes place through hydrolysis and condensation to yield the ceramics condensed around that supramicellar mesostructure. Finally, the surfactant must be removed from the product, which can be carried out through calcination or solvent extraction, leading to a network of cavities within the ceramic framework.

As a consequence of the templating synthetic route, the obtained ceramics will present unique structural properties, such as:

Stable and ordered mesoporous structure
High surface area, in some cases up to ca. 1000 m^2g^{-1}
Large pore volume, up to 1 cm^3g^{-1}
Regular and tunable mesopore size (2–50 nm)
Homogeneous pore morphology

Although the origin of these ordered mesoporous ceramics mainly belongs to the catalysis industry, they have found very promising applications in the biomedical world thanks to their porosity and composition, which promotes their drug delivery capabilities and tissue-engineering potential, as shown in Figure 1.4. Among those, this chapter will introduce calcium phosphates and silica mesoporous materials, since they are the most investigated materials for biomedicine.

The development of these bioceramics produced through conventional methods started in the 1980s. These traditional methods for ceramics production are based in the use of high temperature and pressure. However, these synthetic conditions cannot be employed for biomimetic materials, where the small particle size, in the nanometer scale, is vital. Additionally, the materials obtained through these traditional methods are dense, so they are beyond the scope of this book. However, the fundamentals of these bioceramics will be covered in this chapter because they are the starting point for the production of other types of bioceramics, those bearing porosity, nanometrical particle size, and compositions with applicability in the biomedical world.

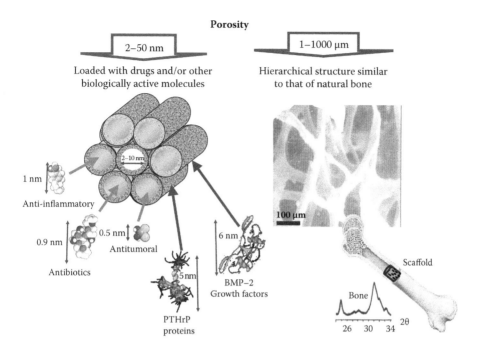

FIGURE 1.4 Porosity ranges of raw mesoporous materials to be employed in biologically active molecules delivery (left) and scaffolds made of mesoporous materials to be employed in bone regenerative technologies (right).

1.2 BIODEGRADABLE AND BIOACTIVE CERAMICS

The main characteristic of bioceramics that are biodegradable and bioactive is that their surface reacts with living tissues through a series of chemical processes to form a mechanically strong bond with these living tissues. This phenomenon of bioactivity will be detailed in the last section of this chapter. These bioactive ceramics are particularly applicable in the clinic when strong bonding between biomaterials and certain living tissues is required (Vallet-Regí 2001). In this section, the most traditionally important resorbable and bioactive ceramics will be reviewed, as can be observed in Figure 1.5.

1.2.1 BIODEGRADABLE CERAMICS

The compositions of the most employed biodegradable ceramics are based on calcium phosphates and calcium sulfates (Elliott 1994; Degroot 1980; Dorozhkin and Epple 2002; Pietrzak and Ronk 2000), which present a similar reactivity degree that leads to positive interactions with living tissues. The main characteristic of biodegradable ceramics is that they are designed to help in the self-repair processes in the living organism for a given period and then resorbed. Thus problems associated with permanence of a synthetic material in the living body will be avoided. The real challenge in the design of biodegradable ceramics is to adapt their degradation kinetics

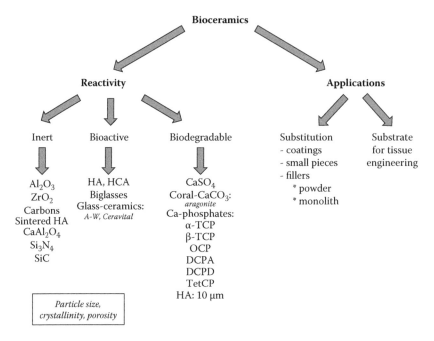

FIGURE 1.5 Different types of bioceramics depending on their reactivity with physiological environments and their applications in biomedicine. HA: hydroxyapatite; HCA: hydroxycarbonate apatite; A-W: apatite-wollastonite; TCP: tricalcium phosphate; OCP: octacalcium phosphate; DCPA: dicalcium phosphate anhydrous; DCPD: dicalcium phosphate dehydrate; TetCP: tetracalcium phosphate monoxide.

to living tissue formation, which is normally slower. Additionally, as resorption takes place, there is a decrease in the mechanical properties of the ceramics, which could be a problem depending on the role assigned to the biodegradable ceramic.

1.2.2 BIOACTIVE CERAMICS

Research on bioactive ceramics exploded in the 1980s thanks to the pioneering work of L. Hench, who was horrified by soldier amputations as a consequence of war. His imaginative idea was based on designing glasses that in contact with living tissue would bond to living bone (Hench et al. 1971). Thus the surface of these bioactive ceramics reacts with the surrounding physiological fluids to form an apatite-like layer that is biologically active. This means that in the presence of living cells it can form novel bone that will join together the ceramic with the osseous tissue. The most representative examples of bioactive ceramics are calcium phosphates, such as hydroxyapatite, and certain compositions of glasses and glass-ceramics (Vallet-Regí 2006; Vallet-Regí, Ragel, and Salinas 2003; Hench and Wilson 1984; H. P. Yuan et al. 1999; Vallet-Regí, Romero et al. 1999; Livingston, Ducheyne, and Garino 2002; Kokubo, Kim, and Kawashita 2003). Table 1.1 shows a selection of traditional bioactive ceramics, which are included

TABLE 1.1

Selection of Traditional Bioactive Ceramics, Different Forms in Which They Have Been Used, and Their Different Applications and Tasks

Material	Form	Application and Task	
Calcium phosphates	Bulk	Bone graft substitutes and cell scaffolds	Replace bone loss
	Coatings	Surface coatings on total joint prostheses	Provide bioactive bonding to bone
Calcium sulfate	Bulk and powder	Bone graft substitutes	Repairing osseous tissues
		Endosseous alveolar ridge maintenance	Space filling and tissue bonding
Glasses	Bulk	Middle ear prostheses	Replacement of part of the ossicular chain
		Orbital floor prostheses	Repair damaged bone supporting eye
		Fixation or revision arthoplasty	Restore bone after prostheses loss
	Powder	Filler in periodontal defects	Periodontal disease treatment
		Bone graft substitutes and cranial repair	Augmentation after diverse illness or traumas
Glass-ceramics	Bulk	Vertebral prostheses	Replace vertebrae removed by surgery
		Iliac crest prostheses	Substitute bone removed for autogenous graft
	Coatings	Fixation of hip prostheses	Provide bioactive bonding

in the second-generation ceramics group, together with their most important clinical applications.

Figure 1.6 represents the different conformations in which bioactive ceramics have been employed in the clinic, such as powder, porous or dense pieces, injectable mixtures, and coatings. Although they show excellent biocompatibility and bioactivity behavior, ceramic materials normally present poor mechanical properties, which limit their applications. In this sense, their brittleness impedes their use for repairing large osseous defects, but they are excellent materials for filling small defects, where the kinetics of bone regeneration is much more important than mechanical properties. Once again in biomaterials technology, the choice of the material would depend on the required application.

Biodegradable and bioactive ceramics can present different crystallization degrees (Figure 1.7), which lead to absolutely different properties. Thus there are crystalline ceramics such as crystalline hydroxyapatite; amorphous solids, such as glasses; and intermediates, such as glass-ceramics, which are formed by an amorphous glassy matrix with crystallization nuclei. In the next section, these bioceramics will be divided by their crystallization stages.

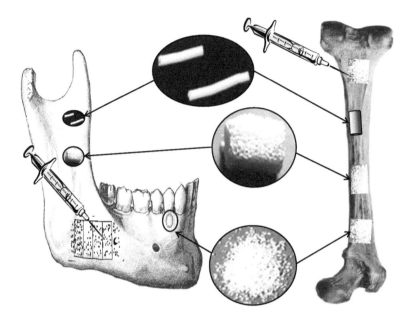

FIGURE 1.6   Different conformations of bioactive ceramics employed in the clinic.

1.3   CALCIUM PHOSPHATES

Calcium phosphates are being used for different applications covering all areas of the skeleton due to their chemical and morphological similarity to the mineral component of bones and teeth. In fact, most of the crystalline biodegradable and bioactive ceramics employed for bone regeneration are based on calcium phosphates. In this section, traditional calcium phosphates will be described, such as synthetic apatites, pure and substituted; biphasic mixtures of calcium phosphates; and calcium bone cements containing calcium phosphates and calcium sulfates. Additionally, the new advances in mesoporous calcium phosphates will also be described.

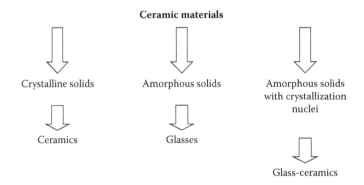

FIGURE 1.7   Different crystallization degrees of biodegradable and bioactive ceramics.

1.3.1 SYNTHETIC APATITES

Apatite refers to a certain group of calcium phosphates, hydroxyapatite, fluorapatite, chlorapatite, and bromapatite, which present high concentrations of OH^-, F^-, Cl^-, or Br^- ions, respectively, in the crystal. From the biomedical point of view, hydroxyapatite, with a chemical formula $Ca_{10}(PO_4)_6(OH)_2$, is the synthetic calcium phosphate most investigated because of the structural and chemical similarities with the inorganic component of bones (Elliott 1994). In this sense, hydroxyapatite with hexagonal symmetry is biocompatible, bioactive, and osteoinductive.

Biological apatites are mimicked in the laboratory to produce synthetic apatites, and to do so, the latter must present nanometrical particle size and should contain between 4 and 8% wt of carbonate anions. Thus synthetic apatites have been obtained in the form of submicrometric particles by aerosol pyrolysis (Vallet-Regí et al. 1994), precipitation (Narasaraju and Phebe 1996; Vallet-Regí, Rodriguez-Lorenzo, and Salinas 1997), and liquid mix technique based on the Pechini patent (Pechini 1967; Pena and Vallet-Regí 2003). Regarding the CO_3^{2-} inclusion in synthetic hydroxyapatites, synthesis at high temperature produced type A (nonbiological) hydroxycarbonate apatite (HCA). When low temperature is employed, type B HCA with very low content of carbonate ions is obtained (Doi et al. 1998). In this sense, during past research it was observed that hydroxyapatite structure can accept many compositional variations, such as the inclusion of many ions, in the Ca^{2+}, PO_4^{3-}, or OH^- sublattices (Zapanta-Legeros 1965; Jha et al. 1997). The included ions, such as Na^+, K^+, Mg^{2+}, Sr^{2+}, Cl^-, and HPO_4^{2-}, can migrate toward living tissues once implanted and promote the increase in the crystal size of these tissues. Thus this migration could have remarkable physiological consequences, since young tissues with low crystallinity can develop and grow faster, at the same time storing other elements that the living body might need during the growing phase.

The properties of certain calcium phosphates can be improved and tailored for specific biomedical applications by hosting ions on their structure. Thus these ionic substitutions can modify the surface charge and the electric charge of hydroxyapatite, which would have a great influence on their performance once implanted. Additionally, the release of some of these elements, such as strontium, zinc, or silicate, during the resorption process of the bioceramic can facilitate the bone regeneration by osteoblasts. Carbonate and strontium ions promote the apatite degradation in physiological environments, leading to the resorption of the implant. Another example of beneficial substitution is the inclusion of silicates, which increase the mechanical properties and accelerate the bioactive response of the hydroxyapatite (Vallet-Regí and Arcos 2005). Thus this great versatility of ion inclusion on hydroxyapatite has shifted the current research trends toward the synthesis of calcium phosphates partially substituted to be used as implantable materials.

1.3.2 BIPHASIC MIXTURES OF CALCIUM PHOSPHATES

The most popular mixtures of calcium phosphates are those formed of hydroxyapatite and β-$Ca_3(PO_4)_2$ (β-tricalcium phosphates, β-TCP), in the so-called β-TCP mixtures (Daculsi 1998; Tancred, McCormack, and Carr 1998; Kivrak and Tas 1998;

Gauthier et al. 2001; Sanchez-Salcedo, Werner, and Vallet-Regí 2008; Sanchez-Salcedo, Arcos, and Vallet-Regí 2008). When these mixtures are in physiological environments, they evolve toward hydroxycarbonate apatite as a consequence of the chemical reactions that take place between the more stable hydroxyapatite and the more resorbable β-TCP. As this mixture gradually dissolves, there is a release of Ca^{2+} and PO_4^{3-} to the local environment, which acts as a stem for newly formed bone.

These types of biphasic mixtures have been found to be very versatile, since they can be employed as injectable materials, as coating of other surfaces, as bulk materials in bone tissue replacement, or as bone defect fillers (Grimandi et al. 1998).

Additionally, there are other biphasic mixtures that have been investigated for biomedical applications, such as other mixtures of calcium phosphates, calcium sulfates (Cabanas, Rodriguez-Lorenzo, and Vallet-Regí 2002), bioactive glasses with hydroxyapatite (Ragel, Vallet-Regí, and Rodriguez-Lorenzo 2002; Ramila et al. 2002), etc.

1.3.3 BONE CEMENTS BASED ON CALCIUM SALTS

The excellent bioactivity and bone-repair properties of cements based on calcium salts, such as calcium phosphates, calcium sulfates, or calcium carbonates, have attracted the attention of biomaterials researchers and physicians in the last few years (Coetzee 1980; Chow 1991; Constantz et al. 1995). The physicochemical properties of these materials, such as setting time, porosity, or mechanical behavior, would depend on the cement formulation, which includes the presence of certain additives to modify those properties as required (Ginebra, Driessens, and Planell 2004; Otsuka et al. 1995; Miyamoto et al. 1998; Nilsson et al. 2002).

These cements are designed to cure in field, and they have been observed to be bioactive with a slow resorption rate. During this gradual degradation, the new bone grows, so the latter would be gradually replacing the former without leaving any empty space. However, these cements need to improve some parameters such as shortening the curing time, improving the mechanical toughness, the application technique on the osseous defect, and some final biological properties.

1.3.4 CALCIUM PHOSPHATE COATINGS

The poor mechanical properties of calcium phosphates, which impede using them in load-bearing applications, motivated the development of coating metallic substrates with those bioceramics. The most employed metallic substrates are pure titanium and titanium alloys, such as Ti6Al4V, because they present low density, great mechanical strength, and resistance to cyclic loads.

The most popular method for the production of commercial implants coated with hydroxyapatite is plasma spray (Koch, Wolke, and Degroot 1990; Sun et al. 2001). The reasons for this popularity are the rapid deposition rate and the low cost. However, there are some synthetic parameters that need to be implemented, such as the presence of amorphous calcium phosphate within the coating, which might promote anomalous resorption rates; the adherence of the ceramic coating to the metallic substrate; the instability of hydroxyapatite at high temperatures; or the phase transition of titanium at 1156 K. Nowadays, there is a great amount of research to improve all these parameters, and it is mainly centered in modifying the synthetic parameters including certain postsynthetic heat treatments to crystallize residual amorphous phases that could be present in the ceramic implant.

Other techniques that require longer times and more investment allow obtaining ceramic coatings on metallic substrates, some of them at lower temperatures, with higher control in the thickness and crystallinity of phases. Some of these techniques are: physical vapor deposition (PVD) (Park et al. 2005), chemical vapor deposition (CVD) (Cabanas and Vallet-Regí 2003), magnetron sputtering (Wolke, de Groot, and Jansen 1998), electrophoretic deposition (Kannan, Balamurugan, and Rajeswari 2002), pulsed laser deposition (PLD) (Arias et al. 2003), and sol-gel-based dip coating (Hijon et al. 2004, 2006). A more detailed review of calcium phosphate coatings for medical implants can be found in a recent book edited by León and Jansen (2009).

1.3.5 MESOPOROUS CALCIUM PHOSPHATE

Calcium phosphates can be manufactured in both porous and dense forms, as well as powders, granulates, or in the form of coatings, as has been mentioned in previous sections of this chapter. Among all calcium phosphates, those with designed and interconnected porosity present very interesting features to be employed in biomedicine, since their interconnected porosity allows transport of body fluids within the bioceramics, enhances their degradation, and increases the possibility of proteins and cells colonizing them (del Real et al. 2003). More concretely, and despite the different methods of promoting porosity using different types of porogens, here we will focus only on mesoporous calcium phosphates produced using surfactants as templates.

The synthesis of these mesoporous bioceramics follows the liquid-crystal templating method described previously in this chapter, that is, the production of ceramics with structural control on the nanometer scale through the interaction with surfactant molecules that can be self-assembled in aqueous solutions (Huang et al. 2011; Prelot and Zemb 2005; Schmidt et al. 2006; Ye, Guo, and Zhang 2008; Ozin et al. 1997; Yuan et al. 2002; Yao et al. 2003; Sadasivan, Khushalani, and Mann 2005; Y. F. Zhao and Ma 2005; Ng et al. 2010). The real challenge in this process is balancing the interactions of phosphate species with calcium ions and amine-type head groups in surfactant molecules simultaneously (Ikawa, Hori et al. 2008). Thus it is very difficult to produce surfactant-templated mesoporous ceramics composed of pure calcium phosphates because calcium and phosphate ions tend to interact strongly, leading to the formation of different crystalline calcium phosphates as by-products, which reduces the total surface area (Ikawa, Oumi et al. 2008; Ikawa et al. 2011). Additionally, it has been found that the surfactant removal in the final step of the synthetic process must be carried out through a washing rather than thermal process to avoid the formation of crystalline phases.

Besides the advantages of calcium phosphates with interconnected porosity that have been mentioned at the beginning of this section, the designed mesoporosity confers on them interesting properties from the drug delivery perspective. Mesoporous calcium phosphates can load a greater cargo of biomolecules than other nonmesoporous calcium phosphates because of their enhanced surface area, and

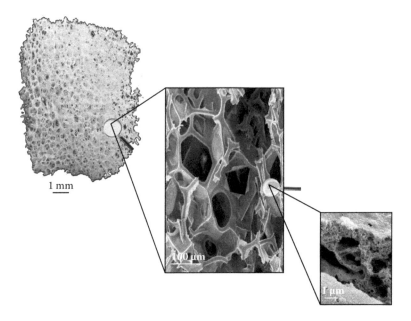

FIGURE 1.8   Digital photograph (left) and scanning electron micrographs (center and right) at different magnifications of hydroxyapatite foams to show the different porosity: macroporous and mesoporous.

they can release the payload with greater control, as has been shown with bovine serum albumin and lysozyme (Ng et al. 2010).

Three-dimensional macroporous hydroxyapatite foams can be produced with a similar approach: the use of the sol-gel process employing surfactants as templates (Sanchez-Salcedo et al. 2010). Thus micro-mesoporous hydroxyapatite with high pore volume and interconnected porosity can be produced. One of the great advantages of this method is the possibility of controlling the macroporosity of the resulting foams by the surfactant:phosphorous source ratio during the synthesis. In this sense, the most homogeneous hydroxyapatite foam obtained through this synthetic method presented an interconnected macroporosity within the range of 1–400 μm, shown in Figure 1.8. What is really interesting about these foams is that their X-ray diffraction (XRD) patterns corresponded to pure hydroxyapatite phase and a porous network within the range of ca. 10–15 nm, that is, a mesoporous network, which could be of interest for further drug delivery applications. Coating the surface of these foams with different biopolymers allows their properties to be modified, such as solubility, mechanical properties, and inducing drug delivery capabilities. This synthetic approach presents great versatility for preparing 3D mesoporous ceramic foams, and their properties could be tuned depending on the biopolymer chosen for the coating.

Three-dimensional interconnected macroporosity of these hydroxyapatite foams allows excellent osteoblast internalization, proliferation, and differentiation, ensuring the cell colonization over the entire material surface (Cicuéndez et al. 2012). Additionally, this mesoporous bioceramic shows an appropriate degradation rate without any cytotoxic effect. This in vitro behavior suggests these biopolymer-coated (gelatine cross-linked with glutaraldehyde and polycaprolactone) nanocrystalline hydroxyapatite macroporous foams as promising candidates as scaffolds for bone tissue regeneration.

1.4 SILICA

Among silica-based bioceramics, the most relevant materials with medical applications are bioactive glasses, which are amorphous materials that were traditionally prepared by the classic quenching of melts comprising SiO_2 and P_2O_5 as network formers and CaO and Na_2O as network modifiers (Vallet-Regí, Ragel, and Salinas 2003). As will be described throughout this section, the production of bioactive glasses has evolved to the use of the sol-gel process, the relatively low synthetic temperature of which allows a greater versatility in composition and structure. Thus it has been possible to combine the traditional composition of glasses with the more modern supramolecular chemistry of surfactants to produce templated glasses, which combine the benefits of bioactive glasses composition together with a mesostructure that offers many possibilities within the biomedical research field. Finally, this section will introduce the pure silica-based ordered mesoporous materials, which were the inspiration for the production of templated glasses, and present many properties very interesting for biomedical technologies.

1.4.1 GLASSES

Glasses can be defined as solid materials with huge structural disorder, or as liquid materials with enormous viscosity values. Thus the main difference between crystalline ceramics and glasses is the order-disorder balance in their respective lattices (Vallet-Regí 2001).

Bioactive glasses were first proposed as implant materials back in 1969 by Hench (Hench et al. 1971). The main characteristic of bioactive glasses is that they are highly reactive and can bond to and integrate with living bone, avoiding the formation of any fibrous capsule around them, and without promoting inflammation or toxicity (Hench 1991). Thus bioactive glasses react with body fluids from the physiological environment to form a layer of hydroxycarbonate apatite nanocrystals, which promotes the generation of bone matrix and accelerates the bone growth process.

Bioactive glasses are used in the clinic mainly for filling osseous cavities, maxillofacial reconstruction, and dental applications (Vallet-Regí, Ragel, and Salinas 2003). Additionally, glasses exhibiting quick bioactive kinetics are sometimes used as bioactivity accelerators of mineral apatites or as bioactivity inductors of magnetic materials for hyperthermia, which have been explored for the treatment of osseous tumors.

Glasses were traditionally prepared by melting processes at high temperature, although in the last few years and thanks to the development in the sol-gel technology, they can be prepared with much greater versatility and at relatively low temperatures through the sol-gel process (Vallet-Regí 2006). Both synthetic processes will be described in the next section of the chapter.

1.4.1.1   Melt Glasses

Melt glasses are produced by quenching the precursor melt prepared at high temperature, which normally contains SiO_2 and P_2O_5 as network formers and CaO and Na_2O as network modifiers. The objective of the network formers is to build Si-O-Si bonds at the same time that the network modifiers break those bonds, which leads to the melt solidification after the quenching with a high degree of disorder. The disordered structure of these glasses provokes a high reactivity when in contact with aqueous solutions, which supposes an advantage when using them in periodontal repair and bone augmentation, since the reaction products of these glasses with physiological fluids lead to crystallization of the apatite-like layer on their surface. The whole process of in vitro bioactivity will be described in the last section of this chapter, but the bioactive mechanism in bioactive glasses is based on the formation of a silica-rich layer on the glass surface when in contact with physiological fluids as a consequence of the ionic interchange of calcium and sodium ions. This layer then attracts calcium, phosphate, and carbonate ions from the physiological solution to form an amorphous calcium phosphate layer, which crystallizes into biologically active hydroxycarbonate apatite nanocrystals. Then this new layer attracts the inorganic moieties to promote the formation of new bone.

The first bioactive glass, produced by Hench as mentioned earlier, was originally within the system Na_2O-CaO-P_2O_5-SiO_2. After this, many different compositions were explored, adding new components, such as K_2O, MgO, CaF_2, Al_2O_3, B_2O_3, or Fe_2O_3, to decrease the temperature of the synthetic process and to improve the properties that were mainly focused on clinical applications. However, it is not always possible to improve all properties simultaneously. The addition of Al_2O_3 aimed toward the improvement of the mechanical properties of the glasses, but it led to the elimination of the bioactivity (Greenspan and Hench 1976). In the same way, the addition of Fe_2O_3 for the production of glasses for hyperthermia treatment of cancer degraded the bioactive behavior of the produced glass (Ebisawa et al. 1997a,b). It was by the end of the 1990s when the compositional limits for in vivo bioactivity within the system Na_2O-K_2O-CaO-MgO-B_2O_3-P_2O_5-SiO_2 were established: 14–30 mol% of alkali metal oxides (Na_2O and K_2O), 14–30 mol% of alkaline earth metal oxides (CaO and MgO), and less than 59 mol% of SiO_2 (Brink 1997; Brink et al. 1997).

1.4.1.2   Sol-Gel Glasses

The sol-gel process is a chemical synthesis method where an oxide network is formed by polymerization reactions of chemical precursors dissolved in a liquid medium. This synthetic method was initially used for the preparation of inorganic materials such as glasses and ceramics at relatively low temperatures. However, this technology has also been developed for the production of organic-inorganic hybrids, as will be detailed later, because the process allows the production of these materials with no thermal degradation of the inorganic content.

The empirical approach of the early studies in the sol-gel process made the technological development grow faster than its own science. Only later, with the advent of modern characterization methods, did the theoretical principles begin to be established; and the published work increased extraordinarily, indicating the interest shown by the scientific community in this promising area.

The development of the sol-gel chemistry at the end of twentieth century brought the first sol-gel bioactive glasses prepared at lower temperatures than melt-derived bioglasses (R. Li, Clark, and Hench 1991). From the biomedical point of view, there are two main advantages of the sol-gel production of bioglasses: it is possible to control the textural properties, which allows porous bioglasses to be obtained (Hench and West 1990); and it is possible to incorporate biological molecules and cells within the matrix network because of the noticeably lower temperatures employed (Avnir et al. 2006; Nieto et al. 2009). Additionally, employing the sol-gel process makes it also possible to increase the range of compositions with bioactive behavior (Vallet-Regí, Arcos, and Perez-Pariente 2000). In general, the sol-gel method is more versatile than the melting-quenching because it allows bioactive coatings, fibers, and highly porous monoliths to be obtained, which can be used as scaffolds for bone tissue engineering. The sol-gel process involves the generation of colloidal suspensions (sols) of solid particles in a liquid. During the process, these colloids are converted to viscous gels (gelation) and then to solid materials.

In the sol-gel process, the most common precursors for the formation of colloidal suspensions are metal alkoxides, which consist of a metal or metalloid element, such as Si, surrounded by various alkoxide groups as ligands. Then hydrolysis and condensation of metal alkoxides form large metal oxide molecules.

1.4.1.2.1 Hydrolysis

Silicon alkoxides are popular precursors because they react quickly with water:

$$Si(OR)_4 + nH_2O \rightarrow Si(OR)_{4-n}(OH)_n + nROH$$

The mechanism of the hydrolysis reaction is thought to occur in three steps. First there is a nucleophilic attack on the silicon atom by the oxygen atom of a water molecule, then there is a transference of a proton from the water to an alkoxide group (OR) of the silicon, and finally there is release of an alcohol molecule (ROH).

Nonsilicate metal alkoxides are very reactive and generally no catalyst is needed for hydrolysis. In the case of silicon-based metal alkoxides, hydrolysis (and also condensation) reactions typically proceed with either an acid or base as catalyst. Depending on the amount and type of catalyst (acid or basic) and also on the amount of water present in the reaction medium, hydrolysis may go to completion or stop, leaving the metal alkoxide partially hydrolyzed:

$$Si(OR)_4 + 4H_2O \rightarrow Si(OH)_4 + 4ROH$$

1.4.2.1.2   Condensation

Depending on experimental conditions (acid or basic catalyzed and condensation of the reactants), two kinds of reactions can take place: water condensation, where two partially hydrolyzed molecules link together liberating a molecule of water:

$$(RO)_3SiOH + HOSi(OR)_3 \rightarrow (RO)_3Si-O-Si(OR)_3 + H_2O$$

or alcohol condensation, where a silanol molecule (partially hydrolyzed silicon alkoxide) condenses with a nonhydrolyzed silicon alkoxide, liberating an alcohol molecule:

$$(RO)_3SiOH + ROSi(OR)_3 \rightarrow (RO)_3Si-O-Si(OR)_3 + ROH$$

Generally, both the hydrolysis and condensation reactions occur simultaneously once the hydrolysis reaction has been initiated.

1.4.2.1.3   Gelation

As the condensation reactions continue, larger and larger silicon-containing polymer networks are formed. Clusters grow by condensation of polymers or aggregation of particles until the clusters collide, and then links form between the clusters to produce a single giant cluster that is called gel. The gel point is the time (or degree of reaction) at which the last bond is formed that completes this giant cluster. Therefore this gel is formed by a continuous solid skeleton that encloses a continuous liquid phase.

1.4.2.1.4   Aging

Segments of the gel network can still move close enough together to allow further condensation, which means that bond formation does not stop at the gel point. Therefore some changes in structure and properties are produced during the aging process. During this process some gels exhibit shrinkage, which occurs because new bonds are being formed where there were formerly only weak interactions between surface hydroxyl and alkoxy groups. This shrinkage leads to expulsion of liquid from the pores of the gels in a process known as syneresis.

1.4.2.1.5   Drying

The loss of water or alcohol from the condensation reactions takes place during the drying process via two different ways: further syneresis (expulsion of the liquid as the gel shrinks) and evaporation of liquid from within the pore structure. This process generates capillary stress, which frequently leads to cracking.

The whole process of the sol-gel processing is summarized in Figure 1.9, together with the application of the sol-gel chemistry to the synthesis of glasses.

The most investigated sol-gel glass composition is that of $CaO-P_2O_5-SiO_2$ (Vallet-Regí, Romero et al. 1999; Vallet-Regí, Arcos, and Perez-Pariente 2000; M. M. Pereira, Clark, and Hench 1994; Peltola et al. 1999; Vallet-Regí, Izquierdo-Barba, and Salinas 1999; Vallet-Regí et al. 2000; Salinas, Vallet-Regí, and Izquierdo-Barba 2001; Balas et al. 2001; Arcos, Greenspan, and Vallet-Regí 2002), although

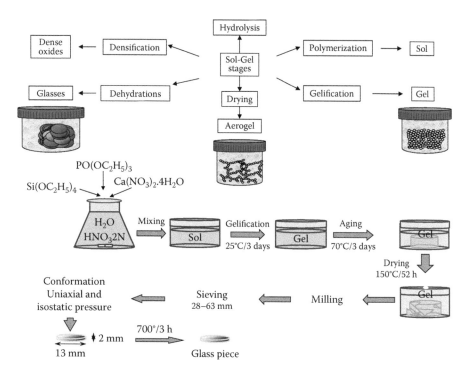

FIGURE 1.9 Schematic representation of the whole sol-gel process (top), and application of sol-gel chemistry for the production of pieces of sol-gel glasses.

it is possible to make some modifications, such as MgO or ZnO additions, or P_2O_5 removal, to modify the mechanical properties, sorption rates, or interaction with osteoblasts (Perez-Pariente et al. 1999; Vallet-Regí, Salinas et al. 1999; Perez-Pariente, Balas, and Vallet-Regí 2000; Du and Chang 2004; Oki et al. 2004; Day and Boccaccini 2005; Du et al. 2006). In terms of bioactive behavior, it has been found that CaO plays an essential role (Martinez, Izquierdo-Barba, and Vallet-Regí 2000; Vallet-Regí et al. 2004), since Ca^{2+} ions leached from the glass increase the Si-OH concentration at the surface of the glass thanks to the interchange with protons from the medium, which favors the formation of the hydroxycarbonate apatite layer, and also increases the Ca^{2+} concentration in the solution, which leads to a supersaturation of the media favoring the apatite-like layer formation. In the same way, the presence or absence of P_2O_5 within bioactive sol-gel glasses has a strong influence on the bioactive behavior, as is described in Figure 1.10 (Salinas, Martin, and Vallet-Regí 2002). In sol-gel glasses produced with P_2O_5, the Ca^{2+} ions are retained in the glass network because of the interaction with phosphate groups, which impedes the media supersaturation in Ca^{2+} and surficial silanol concentration increase, leading to a reduction on the bioactive kinetics. However, once the calcium phosphate phase is formed on the surface of the glass, the crystallization of the hydroxycarbonate apatite is faster thanks to the presence of calcium phosphate (β-TCP) nanocrystals. Transmission electron micrographs of glasses made of $CaO-P_2O_5-SiO_2$ ($G_{Si-Ca-P}$) can

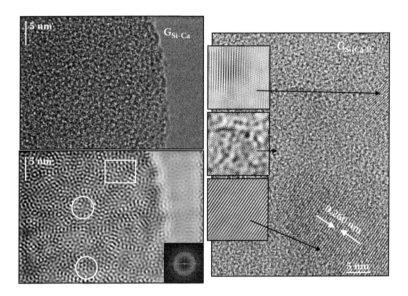

FIGURE 1.10 Transmission electron micrographs of bioactive glasses made of CaO-SiO$_2$ (G$_{Si-Ca}$) and CaO-P$_2$O$_5$-SiO$_2$ (G$_{Si-Ca-P}$). The high-resolution image shows amorphous zones and ordered areas with interplanar spacing of 0.26 nm.

be observed on the right-hand side of Figure 1.10. The presence of P promotes the nucleation of apatite from the very beginning, which is important for the kinetics of the new layer formation and in the size of the produced crystals. On the other hand, P$_2$O$_5$-free sol-gel glasses contain an excess of Ca^{2+} ions within the glass network that can be interchanged with physiological fluids, which leads to the previously mentioned enrichment of the surface with silanol groups and calcium supersaturation of the medium, which favors and accelerates the amorphous calcium phosphate formation, as observed in the electronic transmission micrographs at the left-hand side of Figure 1.10. However, the crystallization of the hydroxycarbonate apatite requires a longer time.

Apart from the bioactive behavior, certain sol-gel glasses have shown excellent biocompatibility in both cell culture studies (Olmo et al. 2003) and animal models (Meseguer-Olmo et al. 2002; Gil-Albarova et al. 2004, 2005). These positive results have inspired and motivated the production of sol-gel glasses as bioactive coatings, mixed materials, and porous scaffolds in third-generation materials, as will be detailed in the following sections of this chapter.

1.4.2.1.6 Bioactive Glass Coatings

As has been detailed previously, there are many different techniques to produce calcium phosphate coatings over metallic substrates when certain mechanical properties are required. In the case of sol-gel bioactive glasses, the advances in the technology of that synthetic approach allow using the dip-coating method for the production of thin films. In fact, this process represents the oldest commercial application of sol-gel technology.

The dip-coating process starts with the production of the precursor solution, whose composition would depend on the type of coating desired. In the case of sol-gel glasses compositions, they are the same precursors as those employed for the synthesis of the bulk materials. Then the metallic substrate is introduced in the sol and withdrawn from the sol at a constant rate. After several seconds, the process becomes steady with the substrate out of the sol. At this point, the entrained film thins by evaporation of solvent and gravitational draining. The microstructure and properties of the film depend on the size and structure of the inorganic sol precursors, the magnitude of the capillary pressure exerted during drying, and the relative rates of condensation and drying. Thus it is possible to vary the porosity, roughness, and composition of the bioactive film obtained, showing again the great versatility that the sol-gel process has to offer.

As mentioned in previous sections, some of the most employed substrates for coating with thin films are titanium alloys because of their low density, great chemical strength, and resistance to cyclic loads. In this sense, coatings of bioactive glass thin films with a composition of 20% CaO-80% SiO_2 (mol%) over Ti_6Al_4V have been successfully produced (Izquierdo-Barba, Asenjo et al. 2003; Izquierdo-Barba, Vallet-Regí et al. 2003). This composition was selected to allow the leaching of calcium ions to the surrounding to favor the bioactive process. When these coatings were soaked in an acellular simulated body fluid, they were completely dissolved after 14 days, because of the high reactivity of these glass compositions. However, when these same thin films were investigated in the presence of osteoblasts, the coatings remained stable and simultaneously promoted an increase of certain cellular parameters, such as adhesion, proliferation, differentiation, or spreading (Izquierdo-Barba et al. 2006).

When coating metallic implants, the interaction between the newly formed sol-gel glass and the metallic substrate itself is something that needs to be carefully considered. That interaction should be strong enough to keep the metallic implant coated with the glass until the hydroxyapatite layer is formed once the materials are implanted. Different studies have been carried out, and when coating sol-gel glasses on 316L stainless steel, the silicon from the glass bonded to chromium from the substrate during the coating process, which increased the corrosion of the metallic alloy (Vallet-Regí, Izquierdo-Barba, and Gil 2003).

1.4.2.1.7   Mixed Materials Containing Bioactive Sol-Gel Glasses
The highly bioactive behavior of certain compositions of sol-gel glasses has fueled the production of many different composite materials with bioactive sol-gel glasses as additives to induce or increase bioactivity. In this section, some examples will be examined, such as: sol-gel glasses with polymers, sol-gel glasses with magnetic materials, and sol-gel glasses with hydroxyapatite.

1.4.2.1.7.1   Sol-Gel Glasses with Polymers   Composite materials made of bioactive sol-gel glasses and polymers were inspired by the possibility of introducing drugs into those polymers and then releasing them once the composite materials are implanted (Arcos et al. 1997; Granado et al. 1997; Vallet-Regí et al. 1997, 1998; del Real, Padilla, and Vallet-Regí 2000; Padilla, del Real, and Vallet-Regí 2002). The advantage of using bioactive glasses is that osseous integration could be improved and the drug release might be favored thanks to the ionic interchange between the glass and the medium.

1.4.2.1.7.2 Sol-Gel Glasses with Magnetic Glass-Ceramics   Sol-gel glasses with magnetic glass-ceramics were initially designed for the treatment of tumor osseous tissues with hyperthermia with the ability of regenerating bone. Hyperthermia treatment is based on the higher sensibility to high temperatures of cancer cells in comparison with healthy cells. Thus body tissue is exposed to high temperatures that can damage and kill cancer cells, usually with minimal injury to normal tissues. In this way, hyperthermia may shrink tumors by killing cancer cells and damaging proteins and structures within cells. Magnetic glass-ceramics might act as thermoseeds to accumulate heat while the sol-gel glass component can induce bioactivity (Ohura et al. 1991; Ebisawa et al. 1991, 1997; Ikenaga et al. 1993; Arcos, del Real, and Vallet-Regí 2002, 2003; Ruiz-Hernández et al. 2006).

However, the incorporation of iron to crystalline phases of glass-ceramics seems to be inhibited somehow by the presence of sol-gel glasses, which reduces the magnetic behavior of these composites (Arcos, del Real, and Vallet-Regí 2003; Ruiz-Hernández et al. 2006). In any case, the in vitro bioactivity of these mixtures has been reported (Serrano et al. 2008), while the in vivo behavior is still under research.

1.4.2.1.7.3 Sol-Gel Glasses with Hydroxyapatite   When the quick bioactive response of sol-gel glasses was observed, it rapidly inspired scientists to employ it as an additive in hydroxyapatite ceramics to accelerate the bioactive response (Oonishi et al. 2000). As was expected, when testing biphasic mixtures of sol-gel glasses and hydroxyapatite higher bioactivity than that of pure hydroxyapatite was observed (Ramila et al. 2002; Vallet-Regí et al. 2003; Padilla, Sanchez-Salcedo, and Vallet-Regí 2005; Padilla et al. 2006). In fact, with only 5% of bioactive glass with a composition of 30% CaO-70% SiO_2, the bioactive response of hydroxyapatite in SBF was noticeably increased from 45 days of immersion for pure hydroxyapatite down to 12 hours for the glass-hydroxyapatite mixture. Additionally, the hydroxycarbonate apatite layers formed on the surface of these composites after the in vitro investigations in physiological fluids were more homogeneous than those formed on pure sol-gel glasses.

1.4.2.1.8 Glass-Ceramics

Glass-ceramics are polycrystalline materials that are produced through a controlled crystallization method and share many properties with both glasses and ceramics. In this sense, glass-ceramics present an amorphous phase, from the glass content, and one or more crystalline phases, from the ceramic part.

The production process of these materials is normally carried out in two stages. First, the glass is formed through a glass-manufacturing process and is cooled down. Then certain nucleating agents, such as Cu, Ag, Au, Pt, TiO_2, ZrO_2, or P_2O_5, are added to help nucleate and grow small crystals uniformly distributed within the glass matrix. Nucleation takes place at lower temperatures than the melting temperature of the glass, so it will not be melted. Finally, further annealing is carried out at the appropriate temperature to promote uniform crystal growth. The control on the grain size, commonly between 0.1 and 1 μm, can be achieved by the careful selection of the nucleating agents and by the heat treatment temperatures.

In terms of bioactivity, glass-ceramics are less reactive at their surface than pure glasses, which means a lower bioactive response. The reason for that is the lower silanol density at the surface of the composite materials and the greater difficulty for calcium ions to be leached out because they are entrapped in the crystalline phases.

When considering these composites as implantable materials for bone tissue, their mechanical properties should be considered. They have been found with better mechanical properties than their parent glasses and sintered crystalline ceramics because of the crystalline phases actuating as reinforced components. However, those mechanical properties of glass-ceramic composites are still lower than those of cortical bone (Vallet-Regí et al. 2005), so more research is needed before there can be a successful commercial application of these materials.

1.4.2.1.9 Ordered Mesoporous Glasses

When the supramolecular chemistry of surfactants was incorporated into the bioglasses field (Figure 1.11), a new generation of sol-gel glasses with tailored porosity at the nanometer scale was developed: the so-called templated glasses or ordered mesoporous bioactive glasses (MBGs) (Yan et al. 2004; Lopez-Noriega et al. 2006).

These materials present the same composition of bioglasses but with designed mesoporosity and enhanced textural parameters, as observed in Figure 1.12. Obviously, those parameters would depend on the template employed. Thus when ionic surfactants, such as cetyltrimethylammonium bromide (CTAB), are employed

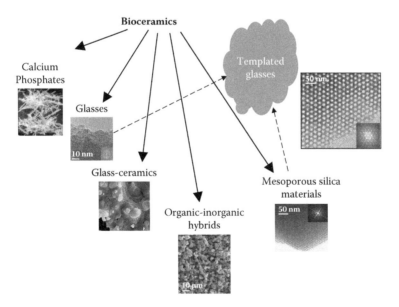

FIGURE 1.11 Schematic representation of the combination of glass composition with the synthetic method of mesoporous silica materials to yield the so-called templated glasses.

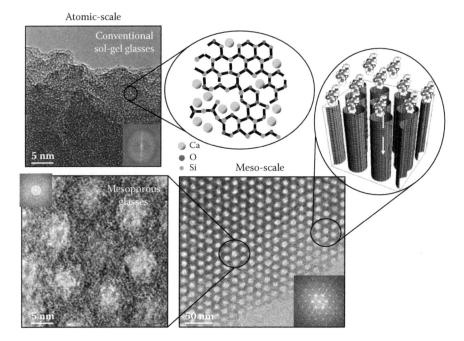

FIGURE 1.12   Transmission electron micrographs and Fourier transforms of conventional sol-gel glasses (top) and mesoporous or templated glasses (bottom). The cartoons represent a scheme of the amorphous structure of traditional glasses and the ordered mesoporous arrangement of templated glasses.

as structure-directing agents, the mesoporous structure is highly defective due to the interference of the surfactant polar head with Ca^{2+} ions. However, when nonionic surfactants, such as triblock copolymers containing poly(ethylene oxide)-poly(propylene oxide)-poly(ethylene oxide) units (EO-PO-EO), are used there is no interference between the ions, and the mesostructure presents a greater order.

MBGs are normally produced employing the evaporation-induced self-assembly (EISA) method (Brinker et al. 1999), which is based on the preferential evaporation of solvent leading to the progressive concentration enrichment of nonvolatile constituents and leading to the ordered mesophase. A schematic representation of the synthesis of three MBGs denoted as Si58m, Si75m, and Si85m is displayed in Figure 1.13 (Lopez-Noriega et al. 2006). The nonionic triblock copolymer Pluronic P123® was used as the structure-directing agent. A constant ratio between the network former precursors and the surfactant was kept. Then, following the steps of the sol-gel process, gelling and drying take place, and the subsequent surfactant removal leads to the formation of MBGs.

The characterization of these materials allowed for the understanding of the influence of the mechanism of synthesis on the chemical composition and textural properties of these MBGs. Figure 1.14 displays N_2 adsorption isotherms for the three MBGs: S58m, S75m, and S85m. All the curves can be identified as type IV isotherms, which are characteristic of porous materials, with H1 hysteresis loops in

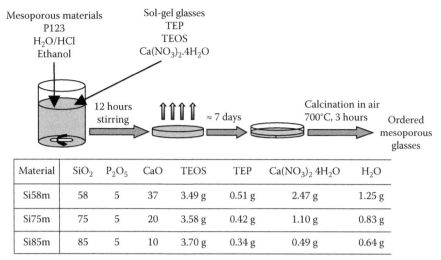

Material	SiO$_2$	P$_2$O$_5$	CaO	TEOS	TEP	Ca(NO$_3$)$_2$ 4H$_2$O	H$_2$O
Si58m	58	5	37	3.49 g	0.51 g	2.47 g	1.25 g
Si75m	75	5	20	3.58 g	0.42 g	1.10 g	0.83 g
Si85m	85	5	10	3.70 g	0.34 g	0.49 g	0.64 g

Nominal composition (mol%) and amounts of reactants (g). All synthesis were carried out with 2 g of P123, 0.5 ml. Of HCl 0.5N and 37 mL of ethanol.

FIGURE 1.13 Schematic representation of the synthesis of MBGs through the EISA method: All the precursors are mixed to then slowly evaporate the solvent to then dry the material for obtaining the templated glass. The table displays different MBGs and the nominal composition (mol%) of reactants used for their synthesis.

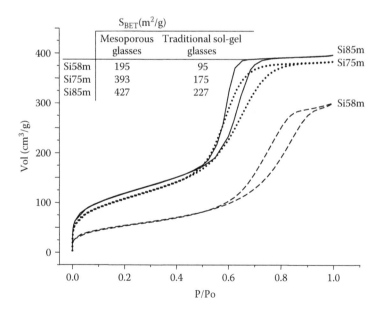

FIGURE 1.14 Nitrogen adsorption-desorption isotherms of three different SiO$_2$-CaO-P$_2$O$_5$ MBGs.

the mesopore range, which are characteristic of cylindrical pores open at both ends. Table inset Figure 1.14 (Lopez-Noriega et al. 2006) shows the surface area (S_{BET}), average mesopore volume, and pore size measured for the three samples, which range between 99 and 227 m^2/g, depending on the SiO_2 content. For comparison, the S_{BET} values of traditional sol-gel glasses also have been included. S_{BET} values of MBGs are significantly higher than those of traditional sol-gel glasses exhibiting analogous compositions (Vallet-Regí, Ragel, and Salinas 2003; Vallet-Regí, Arcos, and Perez-Pariente 2000; Balas et al. 2001). In the same way (results not shown), pore volume is also higher for MBGs compared to that of traditional sol-gel glasses.

Of foremost interest is the possibility of tailoring the structural and textural properties of MBGs depending on the clinical needs (Vallet-Regí, Izquierdo-Barba, and Colilla 2012). As in the case of mesoporous silica materials, the formation of an ordered mesoporous arrangement is driven by different parameters such as surfactant nature and concentration, solvent, use of additives, pH, temperature, etc. (Soler-Illia et al. 2002). Concretely, in MBGs the presence of (1) CaO and P_2O_5 in the silica network as well as (2) solvent evaporation temperature also plays a relevant role in the textural and structural properties of the final matrix.

1.4.2.1.9.1   Influence of CaO and P_2O_5 Content   The CaO content in MBGs is a very important issue, since it not only acts as modifier of the silica network (Balas et al. 2001; Salinas, Vallet-Regí, and Izquierdo-Barba 2001) but also influences the mesopore structure formed after the surfactant removal (Lopez-Noriega et al. 2006). Previous research work demonstrated that decreasing the CaO amount in SiO_2-CaO-P_2O_5 MBGs' porous structure led to an evolution from three-dimensional (3D) cubic structures with *Ia-3d* symmetry to two-dimensional (2D) hexagonal structures with *p6m* symmetry (Lopez-Noriega et al. 2006; Izquierdo et al. 2008), as demonstrated by transmission electron microscopy (TEM) studies (Figure 1.15).

These structural modifications can be explained by the influence of Ca^{2+} ions on the silica condensation, due to the fact that Ca^{2+} ions act as network modifiers and thus decrease the silica network connectivity. As a result, the inorganic:organic volume ratio of the micelle is increased, resulting in a decrease in the curvature radius of the surfactant micelles, which favors the formation of hexagonal phases rather than cubic ones.

The P_2O_5 content also exerts a significant influence on the mesoporous structure. This was demonstrated by synthetizing two MBGs with compositions $85SiO_2$-$10CaO$-$5P_2O_5$ and $90SiO_2$-$10CaO$ (mol%), i.e., in presence and absence of phosphorus, respectively, under the same experimental conditions. The molar ratio between the network formers ($SiO_2 + P_2O_5$) and surfactant was always kept constant (García et al. 2009). TEM studies showed that in the presence of phosphorus, the mesoporous arrangement consisted of a 3D bicontinuous cubic structure, whereas in the absence of phosphorus the structure evolved to 2D hexagonal (Figure 1.16).

To understand this structural change, studies by solid-state nuclear magnetic resonance (NMR) were carried out (Leonova et al. 2008). The 31P solid-state NMR results provided evidence that the presence of CaO and P_2O_5 in $85SiO_2$-$10CaO$-$5P_2O_5$ MBG leads to the formation of amorphous calcium phosphate (ACP) clusters that accumulate onto the material surface (Figure 1.16). Therefore Ca^{2+} ions do not contribute

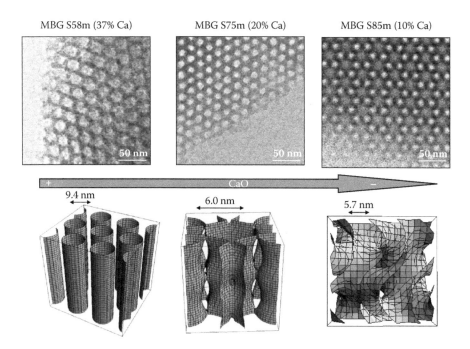

MBG S58m (37% Ca)   MBG S75m (20% Ca)   MBG S85m (10% Ca)

9.4 nm   6.0 nm   5.7 nm

FIGURE 1.15 TEM micrographs of MBGs with different calcium composition (top) and 3D reconstruction of the respective mesostructures (bottom).

to decrease the silica connectivity, and consequently, the inorganic:organic volume ratio of the micelle is decreased, forming cubic structures. On the contrary, in the absence of phosphorus, such as in $90SiO_2$-$10CaO$ MBG, the total calcium content provoked a disruption in the silica network, as confirmed by 29Si-solid state NMR (Gunawidjaja et al. 2012). This can be ascribed to the higher inorganic:organic volume ratio of the micelle, which would increase the curvature radius of the surfactant micelles and contribute to the formation of hexagonal phases rather than cubic ones. Regarding the textural properties, MBGs exhibit higher surface areas and pore volumes than traditional sol-gel glasses owing to the presence of the surfactant during the synthesis of the former, which drive the final porosity characteristics. Therefore although CaO and P_2O_5 contents do not directly control the textural properties of MBGs, they are key parameters that modulate the structure of MBGs, which is in turn associated with well-established textural features.

1.4.2.1.9.2 Influence of Solvent Evaporation Temperature The evaporation temperature used during the EISA process also affects the final mesostructure of MBGs (García et al. 2009). It has been observed that the structure of a MBG evolves from 3D bicontinuous cubic structure to 2D hexagonal structure when decreasing the solvent evaporation temperature. The dependence of a mesoporous structure on the evaporation temperature can be explained in terms of a reduction of hydrogen bond interactions. In the case of nonionic triblock copolymers such as Pluronic P123, the micelle size is strongly dependent on the hydrogen bond interactions with the solvent,

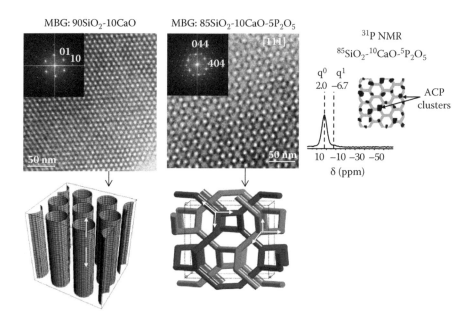

FIGURE 1.16 TEM images and their corresponding of MBGs in the absence or presence of phosphorus. 31P solid-state NMR of P-containing MBG showing the formation of ACP clusters is also displayed. Inset shows ACP distribution into the pore wall of the corresponding P-containing MBG.

which become greater when hydrogen interactions are reduced. Consequently, the hydrophilic:hydrophobic ratio is reduced, favoring hydrophobic mesostructures such as cubic *Ia-3d*, as previously indicated by Zhao and coworkers for pure silica mesoporous materials obtained via the hydrothermal method (Z. Li, Chen et al. 2007).

The textural and structural features of MBGs determine their accelerated in vitro bioactive behavior compared with traditional sol-gel glasses with similar compositions. All the parameters that determine the bioactive behavior of MBGs and their comparison to that of conventional sol-gel glasses will be tackled in Section 1.6.2.3.

As mentioned previously, the great porosity characteristics of mesoporous bioactive glasses can be employed for drug delivery purposes, since drug molecules can be uploaded into the mesopores and then released in a controlled way. Mesoporous materials obtained by conventional templating methods normally exhibit irregular bulk morphology. However, microspheres with homogeneous morphology are widely accepted for drug delivery technologies because they can be ingested or injected into the living body. In this sense, mesoporous bioactive glass microspheres have been observed to present accelerated deposition rates of hydroxycarbonate apatite and great hemostatic efficacy (Ostomel et al. 2006). The same spherical particle morphology has been obtained with mesoporous glasses encapsulating magnetic nanoparticles, which allows these particles to be directed to a target tissue employing an external magnet. In this way, it is possible to accumulate mesoporous glass spheres loaded with a certain drug that can be released specifically at this target place. Additionally, it is also possible to use hyperthermia combined with antitumor

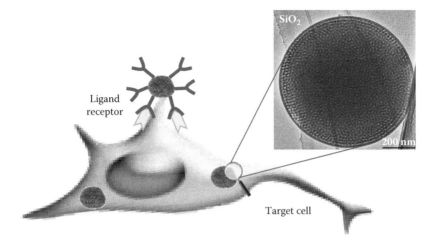

FIGURE 1.17 **(See color insert.)** Schematic representation of mesoporous nanoparticles with targeting agents grafted on their surface to specifically interact with certain receptors to be then internalized to release their cargo.

drugs to treat certain cancerous tissues without affecting healthy tissues (Ruiz-Hernández et al. 2007, 2008). This approach includes the grafting of targeting agents on the surface of mesoporous nanoparticles that could specifically interact with cell receptors to be then internalized in the cell to release their cargo, as can be observed in Figure 1.17.

The degradation products of MBGs in physiological environments are biocompatible, which is one of the most important requirements for implantable materials. In fact, MBGs have shown biocompatible behavior when in contact with osteoblasts, fibroblasts, and lymphocytes (Alcaide et al. 2010).

Chapter 5 will describe some of the recent applications of MBGs aimed at designing 3D scaffolds for bone tissue regeneration purposes. The reader therefore is encouraged to read the chapter for further discussion on this topic (Yun, Kim, and Hyeon 2007; Yun et al. 2007, 2008; X. Li, Wang et al. 2007; Zhu et al. 2008; Yun, Kim, and Hyun 2008, 2009; X. Li et al. 2008; Yun, Kim, and Park 2011; García et al. 2011).

1.4.2 ORGANIC-INORGANIC HYBRIDS

Organic-inorganic hybrid materials represent a relatively novel area of materials science that offers promising solutions to biomedical engineering, such as the ability to bond to osseous tissues or to host molecules with biological activity (Vallet-Regí and Arcos 2006). Hybrid materials experienced a revolution with the development of the sol-gel process, since it allows synthetic paths at relatively low temperatures that would not degrade the organic content of the hybrid, and also allows the production of the materials in different forms, such as bulk, coatings, fibers, etc.

Within implantable materials for bone tissue, organic-inorganic hybrids have been designed to combine the excellent bioactive properties of sol-gel glasses together with the great versatility of organic polymers, which allows improvement of the mechanical properties or drug delivery capabilities, among others. In this sense, the inorganic content of the hybrids is commonly based on the SiO_2-CaO system to ensure the presence of silanol groups at the surface and the leaching of calcium ions to the environment, which are well known to favor the bioactive response. On the other hand, the organic content of the hybrids has traditionally been based on several biocompatible polymers to tailor the hybrid material properties.

A great example of this type of materials is the so-called *star-gel* materials (Manzano et al. 2006), which combine the excellent bioactive properties of sol-gel glasses with mechanical properties comparable to natural bone, which was one of the drawbacks of pure glasses. A scheme of the inspiration for star-gel production can be observed in Figure 1.18.

These star-gels were first developed by the chemical company Dupont back in 1995 (Sharp and Michalczyk 1997), and they present a peculiar structure of an organic core surrounded by flexible arms, as observed in Figure 1.19, which give them their amazing mechanical properties. At the end of these flexible arms there are alkoxysilane groups that, depending on the reaction conditions, can undergo the sol-gel process to form a silica network leading to an organic-inorganic hybrid material.

The mechanical properties of these star-gel materials are derived from their flexibility at the molecular level, which leads to materials with a macroscopic mechanical behavior that is halfway between glasses and rubbers (Sharp 1998). When targeting a bioactive response, calcium ions need to be added to the hybrid network, which makes these materials very attractive from the biomedical point of view. Additionally, when their mechanical properties are evaluated, they have shown fracture toughness comparable to the human tibia. These organic-inorganic hybrids are the only bioceramics produced so far with mechanical properties similar to or even better than human tibia, as is described in Figure 1.20. However, the produced star-gels are dense, and therefore second-generation bioceramics.

In general, there are many different organic-inorganic hybrid materials that have been designed for clinical applications, whether as bioactive or degradable materials. In this sense, polymers with a specific functionality, such as amine groups, can be designed to favor the interaction with biological entities during the bone reconstruction process (Colilla, Salinas, and Vallet-Regí 2006). In this kind of bioactive hybrid materials for bone reconstruction, a common strategy to induce and/or improve bioactive response has been introducing phosphorus or titanium oxide as the inorganic component to the well-known polymers poly(dimethylsiloxane) (PDMS) (Chen et al. 1999, 2000; Manzano, Salinas, and Vallet-Regí 2006; Miyata et al. 2002), or poly(tetramethylene oxide) (PTMO) (Miyata et al. 2002, 2004). Other examples of organic-inorganic hybrids investigated for bone repairing are those whose inorganic content was based on glasses such as CaO-SiO_2 and the organic component was PDMS, obtained as bioactive coatings (Hijon et al. 2005), bioactive monoliths (Salinas et al. 2007), or also doped with P to accelerate the bioactive response (Manzano, Salinas, and Vallet-Regí 2006). Other hybrids for the same purpose include CaO-SiO_2-P_2O_5 as the inorganic content and poly(vinyl alcohol) as the organic part, which have been produced as transparent films (A. P. Pereira,

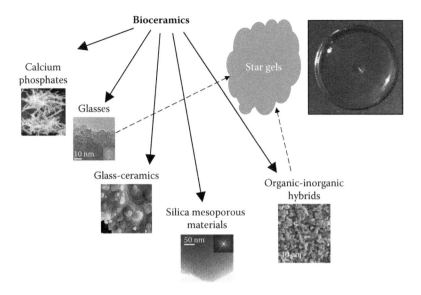

FIGURE 1.18 Schematic representation of the inspiration for the production of star-gels, which are organic-inorganic hybrids with an inorganic content similar to bioglasses.

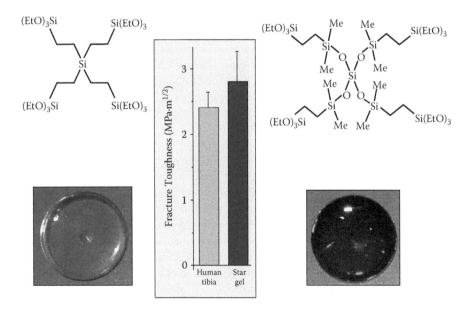

FIGURE 1.19 Chemical structure of the sol-gel precursors (top), the produced monoliths when applying the sol-gel chemistry with TEOS and those precursors (bottom), and fracture toughness values of human tibia and star-gels, which are within the same range.

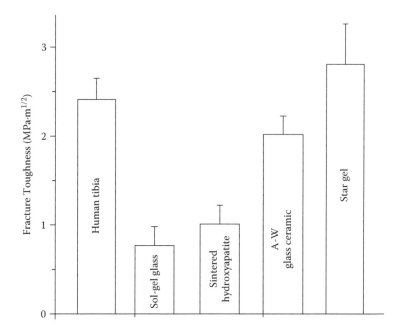

FIGURE 1.20 Fracture toughness of human tibia, different bioglasses, and star-gel hybrid materials.

Vasconcelos, and Orefice 2000) and as degradable monoliths (Martin, Salinas, and Vallet-Regí 2005).

As mentioned earlier, organic-inorganic hybrid materials can also be employed for drug delivery technologies, since polymers were the first proposed materials for drug release as degradation would take place. In this sense, it is possible to combine the bioactive properties of glasses such as $CaO-SiO_2$ with the drug delivery capability of certain methacryloxy and amino polymers (Colilla, Salinas, and Vallet-Regí 2006) to produce hybrid materials for time release of certain bioencapsulates (Gonzalez, Colilla, and Vallet-Regí 2008).

The great versatility in the synthetic process of organic-inorganic hybrids has allowed the production of porous scaffolds for tissue engineering (Tsuru, Hayakawa, and Osaka 2008). One of the most important features of these porous scaffolds is that their porosity should be interconnected to allow the physiological fluids go through and cells to colonize them. This approach was followed for the synthesis of scaffolds made of SiO_2-PDMS with 90% of porosity and pore size ranging between 200 and 500 µm, in which the porosity was obtained using sucrose particles as template (Yabuta et al. 2003). These scaffolds have been implanted into brain defects with quite positive output. Another type of scaffold can be prepared using gelatin-siloxane to obtain several orders of porosity (5–10, 30–50, and 300–500 µm), which may be interesting for multifunctional applications. They were produced by freeze-drying the wet precursors during the gel stage of the sol-gel process, and have been tested in brain defects without producing any inflammation (Deguchi et al. 2006). The same synthetic procedure of freeze-drying a gel can be employed to produce scaffolds with chitosan-gelatin siloxane bearing 90% of porosity and pore size of ca. 100 µm (Shirosaki et al. 2008).

Organic-inorganic hybrid materials still offer a great range of possibilities that are very interesting for bone tissue regeneration applications, and this is the reason why there is a lot of research being conducted nowadays in this particular area.

1.4.3 PURE SILICA ORDERED MESOPOROUS MATERIALS

The roots of all mesoporous materials are within pure silica ordered mesoporous materials, which were developed in the early 1990s, as has been mentioned previously. These materials are composed of SiO_2 forming Si-O-Si bonds and Si-OH at the surface. This class of materials, characterized by large surface areas, ordered pore systems, and well-defined pore size, needs the supramolecular chemistry of surfactants as structure-directing agents to be produced.

1.4.3.1 Synthesis of Silica-Based Ordered Mesoporous Materials

The formation process of these materials can be explained by two different mechanisms (Figure 1.21):

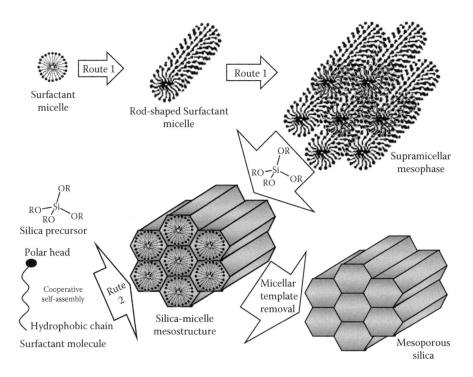

FIGURE 1.21 Two different mechanisms proposed for the production of silica mesoporous materials. Observe that the main difference occurs when adding the silica precursor (TEOS). After the surfactant has formed, the supramicellar phase is formed (route 1), or at the same time as the micellar mesophase is being formed (route 2).

1. The so-called true-liquid-crystal template (TLCT) mechanism, in which the surfactant concentration is so high that a lyotropic liquid-crystalline phase is formed without the need of any inorganic precursor, normally alkoxysilanes. The employed surfactants are amphiphilic substances with hydrophobic groups (their tails) and hydrophilic groups (their heads). Thus a surfactant molecule contains both a water-insoluble (or oil-soluble) and a water-soluble component. The surfactant molecules are dissolved in an aqueous solution at a concentration above the critical micellar concentration (cmc), so the surfactant molecules would aggregate themselves to form micelles. In these micelles, the hydrophobic tails form the core of the aggregate due to unfavorable interactions with water, and the hydrophobic heads are in contact with the aqueous phase. The shape and size of the aggregates would depend on the chemical structure of the employed surfactant, especially on the balance of the sizes of the hydrophobic tail and hydrophilic head, in the so-called hydrophilic-lipophilic balance (HLB). Additionally, the synthesis conditions, such as surfactant concentration, temperature, pH, and additives such as cosurfactants, will also play an important role in the final size and/or shape of the micelles. In a following step, these micelles aggregate themselves to form supramicellar structures that can present different geometries, such as hexagonal, cubic, or laminar, depending on the synthetic conditions. This is a key point in the whole process, since the geometry of the supramicellar aggregates will determine the geometry of the resulting mesoporous framework. At this point is when the inorganic precursor, normally tetraethyl orthosilicate (TEOS) or tetramethylorthosilicate (TMOS), is added to the template and the sol-gel chemistry begins to take place. Hydrolysis followed by condensation of the precursor leads to the formation of a silica gel around the supramicellar aggregates. Further aging and drying yield the silica material (still with the surfactant molecules present) with an ordered mesoporous framework. The final step of the synthetic process is the surfactant removal, which can be carried out through thermal treatment (calcination).

2. The second proposed mechanism is the so-called cooperative liquid-crystal template (CLCT), in which the surfactant concentration is below the cmc and the lyotropic liquid-crystalline phase is formed by a somehow cooperative self-assembly of the structure-directing agent and the already added inorganic precursors. Once the lyotropic liquid-crystalline phase is formed and the alkoxysilane precursors are present in the reaction media, the sol-gel process would undergo the same process as was explained for the TLCT method, leading to ordered mesostructured materials with hexagonal, cubic, or laminar arrangement. As in the previous mechanism, the final step of the synthetic process is the surfactant removal, which can be achieved by thermal treatment (calcination) or solvent extraction. Both methods are equally valid for removing the structure-directing agent from the material, and the choice will depend on the composition and the final application of the material. In any case,

the surfactant removal will lead to a network of cavities within the silica framework that determines the physicochemical characteristics of the obtained materials.

Whatever the mechanism proposed to obtain mesoporous materials, a common requirement is the need of an attractive interaction between the hydrophilic heads of the surfactant template and the silica precursor to avoid phase separation between them. This is a very difficult condition to fulfill when producing mesoporous calcium phosphates, as stated in the previous section, but an easy task to achieve with silica precursors. These interactions would be determined by the type of surfactant employed, such as ionic (electrostatic interactions) or nonionic (hydrogen bonds interactions) surfactants (S), and synthesis conditions, which determine the reactivity of the inorganic precursor (I), and they can be classified as follows (Figure 1.22):

S⁺I⁻: This type of interaction takes place with cationic quaternary ammonium surfactants (S⁺) and when the reaction takes place under basic conditions, it creates silica species anions (I⁻).

S⁺X⁻I⁺: This interaction takes place when employing cationic surfactants under acidic conditions (below the isoelectric point of silica) where silica species are positively charged. In this particular case, to ensure the surfactant-silica interaction it is necessary to add a mediator anion (X⁻).

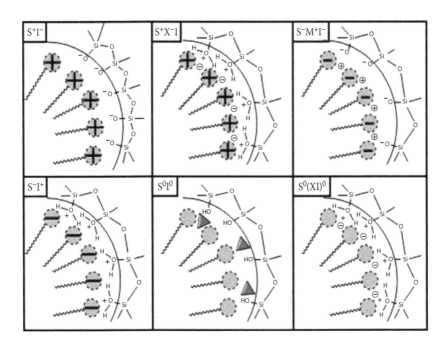

FIGURE 1.22 Graphical description of the six possible interactions between the surfactant (S) and the inorganic precursor (I).

S⁻X⁺I⁻: In this case, the surfactant used is negatively charged (S⁻), such as long-chain alkyl phosphates, and the reaction takes place in basic media, so the silica species are anions (I⁻). As in the previous case, a cation mediator (M⁺) is necessary to ensure the interaction between the negatively charged surfactant and the negatively charged species.

S⁻I⁺: When the surfactant employed is negatively charged (S⁻) and the reaction is carried out in acidic media (I⁺), there is no need for any mediator to ensure the surfactant-silica interaction.

S⁰I⁰: This type of interaction takes place between nonionic surfactants, such as triblock copolymers like EO-PO-EO, with uncharged silica species.

S⁰(XI)⁰: In this case, the interaction takes place between nonionic surfactant and charged silica species that are paired normally with halides.

Taking into account all these different templating possibilities, there are many different mesoporous structures that have been produced in the last few years. Among the most representative are MCM-n (Mobil Corporation Matter) materials (Kresge et al. 1992; Beck et al. 1992; Firouzi et al. 1997; Kaneda et al. 2002; Zhang, Luz, and Goldfarb 1997), SBA-n (Santa Barbara) series (Ravikovitch and Neimark 2002; D. Y. Zhao et al. 1998), MSU-n (Michigan State University) materials (Bagshaw, Prouzet, and Pinnavaia 1995), KIT-n (Korean Institute of Technology) series (Ryoo et al. 1996), FSM-n (Folded Sheet Materials) series (Inagaki et al. 1996), FDU (Fudan University) materials (Fan et al. 2003; Meng et al. 2005; Yu, Yu, and Zhao 2000), and AMS-n (Anionic Mesoporous Silica) series (Che et al. 2003).

1.4.3.2 Characterization of Silica-Based Ordered Mesoporous Materials

A reliable characterization of the ordered porous structure requires the sum of three independent techniques: X-ray diffraction (XRD), transmission electron microscopy (TEM), and nitrogen adsorption analysis (Figure 1.23).

The XRD patterns of these mesoporous materials show reflections only at low angles, because taking into account that these materials are not crystalline at the atomic level, no reflections at higher angles are expected. In the case of MCM-41, the first silica mesoporous material produced and probably the most studied, the mesostructure presents a honeycomb-like morphology, which is the result of hexagonal packing of unidimensional cylindrical pores. The MCM-41 XRD pattern shows typically three to five reflections between $2\theta = 2°$ and $5°$. The reflections are produced because of the ordered hexagonal array of parallel silica tubes and can be indexed assuming a hexagonal unit cell as (100), (110), (200), (210), and (300). In some situations, the intensity of XRD reflections might be small, which means a decreased domain size, but the material is still ordered. What is more important is that depending on the arrangement of the mesopores, that is, on the type of mesostructure, a different diffraction pattern would be obtained. The diffraction patterns of the different types of ordered mesoporous materials have been already described and published.

TEM is usually employed to elucidate the pore structure of these mesoporous materials. With this microscopy technique it is possible to have an idea of the pore size and the thickness of the pore walls, although additional simulations are needed because of the focus problem.

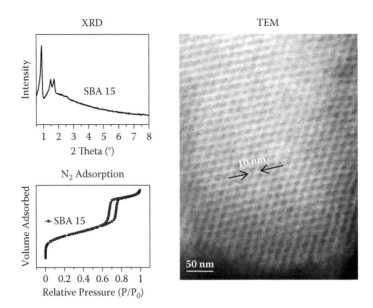

FIGURE 1.23 Different techniques for the characterization of ordered mesoporous silicas. Top left: small-angle X-ray diffraction pattern of SBA-15; bottom left: nitrogen adsorption/ desorption isotherms of SBA-15; right: transmission electron micrograph of SBA-15.

Adsorption analyses, normally of nitrogen molecules, have been employed to determine the surface area and the pore size distribution of ordered mesoporous materials. In the case of MCM-41, the nitrogen adsorption isotherm is type IV in the IUPAC classification, showing two different areas: a sharp capillary condensation step at a relative pressure of 0.4 and no hysteresis between the adsorption and desorption branches. Depending on the type of the mesostructure, there are several methods based on geometrical considerations to determine the pore-size distributions.

The original synthesis of MCM-41 materials pioneered this area of mesoporous materials, but during this time little or no attention was put on the morphology of the produced materials. In these first mesoporous materials, aggregates and loose aggregates of small particles were observed. However, thanks to the advances in this area and the motivation for many applications, it is now possible to obtain defined morphologies of mesoporous materials, such as thin films, fibers, spheres, and monoliths (Ciesla and Schuth 1999). Hollow or hard spheres of mesoporous silica are of great interest for the biomedical research area for the preparation of drug delivery nanoparticles, and in the next chapters of this book the synthetic process and applications will be detailed.

1.5 ORGANIC FUNCTIONALIZATION OF ORDERED MESOPOROUS CERAMICS

The possibility of combining the features of organic and inorganic chemistry in a single material is particularly attractive for materials science because it allows the combination of the wide functional variation of organic chemistry with the advantages of a thermally stable and robust inorganic substrate. This symbiosis of organic and inorganic components permits obtaining materials whose properties are totally different in comparison to those of their individual components, and is of particular interest for heterogeneous catalysis, chromatography, and biomedicine.

The organic functionalization of ceramic materials implies the modification of the chemistry at the surface of the materials, which is a milestone in biomaterials science, since it allows customizing their properties and tailoring their response to physiological environments. Additionally, in the case of ordered mesoporous ceramics, it also allows the modulation of the adsorption and release properties when designing drug delivery devices.

The organic modification has acquired particular notoriety in the case of mesoporous bioactive glasses and ordered mesoporous materials, mainly because of the simplicity of the process and the outcome benefits for drug delivery technologies. There are two synthetic approaches for the production of inorganic mesoporous materials with organic functionalities within their structure: (1) grafting of organic functionality, which consists of the modification of the material surface of a pure silica material; and (2) co-condensation with an organic functionality, which consists of the simultaneous condensation of the silica precursors and the organic functionalizing moieties (Hoffmann et al. 2006).

1.5.1   Post-synthesis Functionalization Method

This synthetic method is based on the fact that the surface ordered mesoporous glasses and silica are all covered by silanol groups (Si-OH). This high density of silanol groups is susceptible to organic modification by covalently grafting diverse functionalities through a reaction with organosilanes, $(R'O)_3SiR$; chlorosilanes, $ClSiR_3$; or silazanes, $HN(SiR_3)_3$, as depicted in Figure 1.24. This is the reason why this functionalization method is also known as grafting.

There are a great number of different organic groups (Figure 1.25) that can be grafted to the surface of the mesopores through this functionalization method by the variation of the organic residue R on the precursors.

The main advantage of this functionalization method, apart from the previously mentioned organic variability, is that the mesostructure of the starting material is normally retained, although there might be a reduction in the porosity due to the presence of the organic moiety.

However, one of the disadvantages of this method is that some organosilanes tend to react preferentially at the pore entrance, which might impede the entrance of further organosilanes to the pores. This could lead to a nonhomogeneous distribution of the organic groups within the walls of the pores and, in some cases, block the entrance of the pores.

1.5.2   Co-condensation Method

This is an alternative functionalization method of ordered mesoporous silica that is based on the co-condensation, during the synthetic process, of the silica precursors together

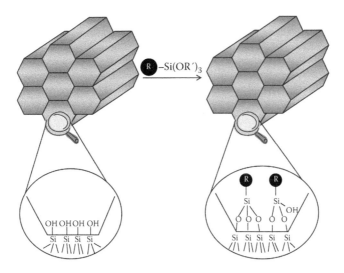

FIGURE 1.24 Schematic representation of the postsynthesis functionalization method of ordered mesoporous materials.

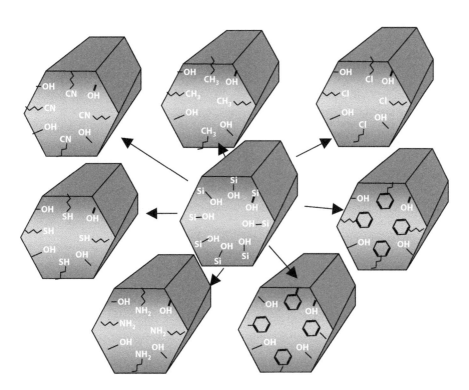

FIGURE 1.25 Schematic representation of the different organic groups that can be grafted to the surface of the pore channels of ordered mesoporous materials.

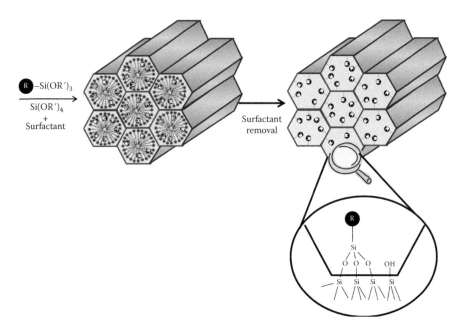

FIGURE 1.26 Schematic representation of the co-condensation functionalization method of ordered mesoporous materials.

with the selected organosilanes. Also, mesostructured silica materials can be prepared by the co-condensation of tetra-alkoxysilanes, $(RO)_4Si$, such as TEOS or TMOS, with trialkoxysilanes, $(RO)_3SiR'$, as described in Figure 1.26. Obviously, this co-condensation step should be carried out in the presence of structure-directing agents, so ordered mesoporous materials with organic residues anchored to the pore walls will be obtained. Special care should be taken to avoid destroying or degrading the organic functionalization during the surfactant removal. This is the reason why calcinations should not be used, and it is preferable to use extractive methods with certain solvents.

This functionalization method is compatible with most of the templating agents utilized for the preparation of the most popular mesoporous silica materials, such as the MCM or SBA series.

The main advantage of the co-condensation method in comparison with the grafting method is that the organic functionalities are generally more homogeneously distributed along the pore walls. In fact, there is no risk of pore blocking, because the organic units are direct components of the silica matrix.

However, as happened with the grafting method, there are also some disadvantages of the co-condensation method. Perhaps the most important limitation is that when increasing the amount of organosilanes in the reaction mixture, the degree of mesoscopic order decreases, which could even lead to totally disordered products. Additionally, it has been observed that when increasing the organic functionalities, there was a reduction in the pore diameter, pore volume, and specific surface area. Finally, the use of different silica precursors implies different hydrolysis and condensation rates of the reactants, which favors the homocondensation reactions rather than heterocondensation. Thus there is a lower yield in the incorporation of organic functionalities to the walls than expected, because the organosilane reacts with itself instead of cocondensing with the silica precursor.

1.6   IN VITRO STUDIES

In this last section of the chapter, the different roads that the scientific community has been exploring to determine the possible bioactivity of a ceramic material will be described. Nowadays, it seems quite reasonable that a material should undergo different characterization methods before being implanted into an individual, or even before using on animals, where ethical aspects associated with animal suffering should be taken into account.

However, back in 1972, when the first bioactive glass produced in the lab was directly implanted in experimental animals (Hench et al. 1971), an interesting output came through that undoubtedly marked the line on the development of bioactive ceramics. After glass implantation in the animals, it was observed that between the glass surface and the natural bone there was a new bonding layer of hydroxycarbonate apatite nanocrystals similar to that present in bone. This new layer at the interface would serve as reinforcement of collagen fibers during the bone regeneration process. Thus the authors of that time considered that a material was bioactive when it was coated by a layer of hydroxycarbonate apatite nanocrystals, as represented in Figure 1.27, when soaked in a fluid-simulating blood plasma.

However, this is just the first trial that a biomaterial should fulfill before being used as bone implant. It is quite obvious that the reasonable path for evaluating a potential implantable material should start by testing it with acellular solutions, but then it also needs to be tested with different cell cultures, to find out possible cytotoxicity, with animal models, to find out the biocompatibility, and finally with humans, following all the clinical trials normative for each country. This path is represented in Figure 1.28.

However, implementing the in vitro testing of biomaterials can save lots of money and resources for the different research agencies since it is much cheaper and easier than in vitro tests with cells, and in vivo tests with animals and patients. It is for this reason that there is a whole section on the in vitro tests of bioceramics within this chapter.

1.6.1   IN VITRO TEST WITH BUFFERED AQUEOUS SOLUTION

When glasses were starting to be produced as implants back in the 1970s and 1980s, the trend was to evaluate them in vitro just in aqueous solution buffered at pH 7.4 (R. Li, Clark, and Hench 1991). The reason was that the investigated glasses were very reactive and already contained calcium and phosphate ions to be released to the medium, which were thought to be enough to form the layer of hydroxycarbonate apatite layer in case the glasses were bioactive.

1.6.2   IN VITRO TEST WITH SIMULATED BODY FLUID

However, by the end of the 1980s, Kokubo and coworkers realized that glass-ceramics that had the ability to bond to living bone in vivo did not form the hydroxycarbonate

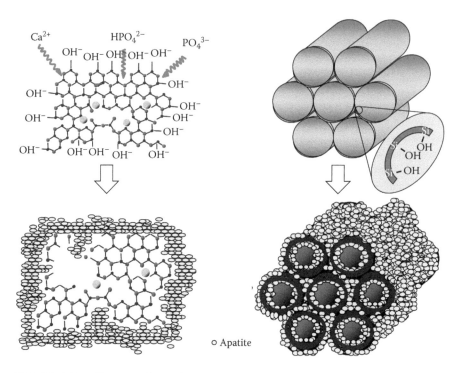

o Apatite

FIGURE 1.27 **(See color insert.)** Representation of the formation of a new apatite-like layer on the surface of glasses (left) and ordered mesoporous materials (right) when in contact with physiological fluids.

apatite layer when tested in the previously mentioned buffered aqueous solution (Kokubo et al. 1990). These findings inspired them to produce a more elaborate solution that mimicked human plasma, the so-called simulated body fluid (SBF). SBF is an acellular ionic solution with an inorganic composition almost equal to human plasma and buffered at physiological pH, as can be observed in Table 1.2.

The difference between the original SBF and the corrected SBF version is the presence of sulfate ions at the same concentration as in plasma (Kokubo 1991). Even though there have been many additional modifications to these compositions, soaking the tested material in SBF under static conditions is a cheap, easy, and quick method to evaluate the possible bioactivity. In fact, when comparing this method with traditional in vivo tests where the bioactivity of a material when in contact with living tissue is evaluated after weeks or months, in the SBF in vitro test materials can be assessed within minutes of immersion. This method has been accepted by the scientific community as the starting point in the evaluation of a material for bone tissue regeneration.

The SBF test has allowed for the determination of the mechanism of bioactivity and for the subsequent development of new bioactive glasses and glass-ceramics. This bioactive mechanism is a surface phenomenon where the presence and concentration of silanol groups at the glass surface has a strong influence when soaked in SBF. It was found that the higher the surface and silanol concentration, the quicker

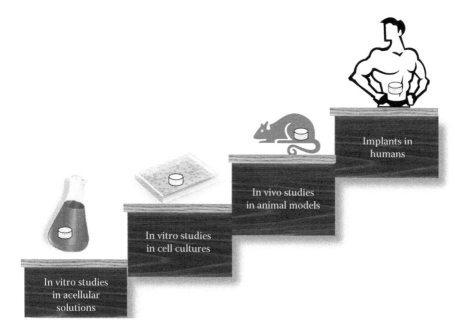

FIGURE 1.28 Schematic representation of the reasonable path for evaluating a potential implantable material.

TABLE 1.2
Ionic Compositions of the Different Solutions Employed for Testing Potential Implantable Materials

	Na⁺	K⁺	Mg²⁺	Ca²⁺	Cl⁻	HCO₃⁻	HPO₄²⁻	SO₄²⁻	pH	Buffer
Blood plasma	142	5	1.5	2.5	103	27	1	0.5	7.35–7.43	Biological
Buffered solution	—	—	—	—	—	—	—	—	7.4	Tris*
SBF**	142	5	1.5	2.5	148.8	4.2	1	—	7.4	Tris
Corrected SBF	142	5	1.5	2.5	147.8	4.2	1	0.5	7.4	Tris
CSIP***	142	5	1.5	2.5	103	24–27	1	0.5	7.4	CO₂/HCO₃⁻

Note: Concentrations are expressed in nM.

* Tris buffer = tris(hydroxymethyl)aminomethane and hydrochloric acid

** SBF = simulated body fluid

***CSIP = carbonated simulated inorganic plasma

the bioactive response. This invaluable information was taken into consideration to design porous sol-gel glasses with high surface area and high silanol concentration (R. Li, Clark, and Hench 1991). Certain thermal treatments and the relative proportion of CaO to SiO_2 were revealed as key factors to obtaining those porous glasses with favorable bioactive behavior (Vallet-Regí et al. 2004). However, the actual revolution in terms of bioactive kinetics came with the development of template glasses, which have been described in a previous section of this chapter, because of the outstanding values of surface area and great mesoporosity, both characteristics that promote a greater reaction degree between the glass and the physiological fluid (Lopez-Noriega et al. 2006).

1.6.2.1 Parameters of in Vitro Bioactivity Tests

In this section the different parameters that are possible to modify during the in vitro bioactivity tests will be detailed. They include the different composition of the acellular solutions employed to soak the material, the static or dynamic conditions employed, or the possible shape of the materials.

As mentioned earlier, the first acellular solution to evaluate the potential bioactivity of certain glasses was an aqueous solution buffered at pH 7.4 with tris(hydroxymethyl)aminomethane and hydrochloric acid. The tested glasses contained Ca, P, Si, and Na oxides. The Ca^{2+} ions were quickly leached to the acellular solution, which promoted the formation of the hydroxycarbonate apatite layer on the surface of the glass (Hench 1991).

However, when the same buffered solution was employed to evaluate different ceramics, such as apatite-wollastonite (A-W) glass-ceramic, the layer of hydroxycarbonate apatite was not formed. These findings confirmed that the solution needed calcium and carbonate ions, since it was not enough to produce the new layer with the ions leached from the glass. This was the motivation to produce a new acellular solution, the so-called simulated body fluid (Kokubo et al. 1990), as was mentioned before in this chapter. SBF is similar to the previous buffered aqueous solution in that it also uses tris(hydroxymethyl)aminomethane and hydrochloric acid to buffer the pH at 7.4, but it differs in the presence of several inorganic ions, whose concentration is almost equal to human plasma. The difference in SBF and acellular plasma comes with the concentration of carbonate ions, 4.2 nM in SBF and 27 nM in human plasma, because it is not possible to reach higher carbonate concentration at ambient conditions. The carbonate anion deficit in SBF is offset by the presence of greater concentration of chloride ions, so the electroneutrality will be ensured. Attending to the ionic concentrations, it is obvious that SBF is a supersaturated solution, so during its preparation special care needs to be taken to avoid any undesired precipitation. There is a very interesting scientific publication where Kokubo details all the steps that need to be taken to successfully prepare SBF (Kokubo and Takadama 2006).

However, SBF is a metastable solution so it must be prepared just before being used. Although it can be stored refrigerated, this can be done for only a limited period of time (no more than one month). This is one of the main drawbacks of SBF, and for this reason there are some authors who have ed different alternatives. One of them was trying to avoid using tris since it is not a physiological buffer and also tends to complex calcium ions, which could influence the bioactive mechanism.

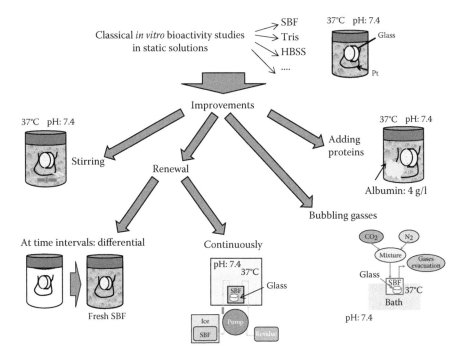

FIGURE 1.29 Different experimental approaches for evaluating the in vitro bioactivity of bioceramics.

Thus tris buffer can be substituted by a more biological friendly CO_2/HCO_3^- buffer to produce the so-called carbonated simulated inorganic plasma (CSIP) (Marques et al. 2004). Other authors have also proposed using commercial solutions that are stable for longer periods of time than SBF, such as Hanks' balanced salt solution (HBSS). Figure 1.29 depicts the different approaches that have been taken in the in vitro bioactivity tests.

During the formation of the hydroxycarbonate apatite, as in all chemical process, the stirring conditions represent an important factor to take into consideration when analyzing the kinetics of the process. Thus the bioactivity test can be done on static conditions or under different levels of agitation. Stirring is necessary to ensure the homogenization of the dissolution, thus facilitating the reaction to take place. However, excessive stirring can influence the deposition of the new phase on the ceramic surface, hindering it. Thus the best way to carry out these bioactive tests is by orbital stirring, rather than the more vigorous magnetic or mechanical stirring.

The current conformation methods allow obtaining materials with the same chemical composition but with different shapes, such as powder, pieces, compacted discs, etc. On one side, there is a trend that defends the following thesis. Bioactivity is a surface process, so this surface should be increased as much as possible to favor this process. Thus these authors recommend performing the in vitro bioactivity tests using samples as powders, so there would be more surface contact of the solid with the solution, thus increasing the subsequent reactivity. On the other hand, there is

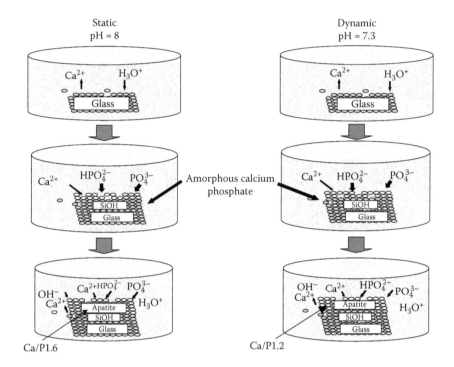

FIGURE 1.30 Different mechanisms proposed for the in vitro formation of the new hydroxycarbonate apatite layer under static (left) and dynamic (right) conditions.

another trend that defends the idea of using samples as pieces, so there would be a relatively large sample surface where it would be possible to apply different characterization techniques to find out about the SBF bioactivity test. If the obtained material is a powder, it is possible to compress it using uniaxial and/or isosthatic pressure.

During the evaluation of certain glasses' bioactivity behavior, it was found that there were significant changes in the concentration of the different species in the dissolution as a consequence of the new layer formation. In this sense, some authors have proposed the renewal of the solution, which can be made at intervals (differential method), or continuous renewal, which can be made using peristaltic pumps (dynamic method) (Izquierdo-Barba, Salinas, and Vallet-Regí 2000).

1.6.2.2 Techniques to Verify Bioactive Response

The majority of the characterization techniques for in vitro bioactivity are based on the analysis of the new layer formed on the surface of the bioactive ceramics. Taking into account that the crystal size of this new layer is very small at the beginning of the process, techniques, such as X-ray diffraction (XRD) or electron diffraction (ED), are not very useful.

However, Fourier transform infrared (FTIR) spectroscopy has been revealed as a powerful technique, since it is very sensitive to the presence of small amounts of phosphate groups. Additionally, FTIR also can detect the presence of silanol groups on the surface of the glass, which play an important role in the bioactive process,

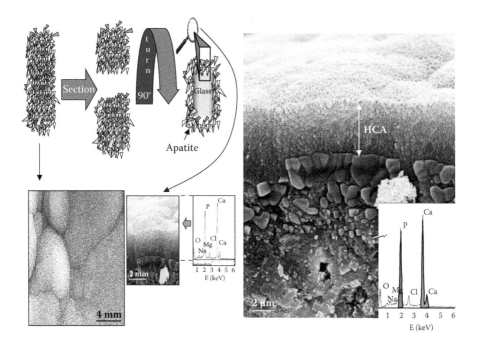

FIGURE 1.31 Scanning electron micrographs employed to evaluate the bioactive response on the surface of the tested bioceramics after being soaked in physiological solutions.

as has been described previously. When hydroxycarbonate apatite is being formed, FTIR allows for the detection of the carbonate groups and also of a certain degree of crystallinity in the calcium phosphate formed.

A powerful microscopic technique such as scanning electron microscopy (SEM) is also very valuable for characterizing the new layer formed as a consequence of the bioactive process. If a backscattered electron detector is used, it is also possible to obtain information on the composition of the latter in terms of atomic percentage (Figure 1.31).

All these characterization techniques shown in Figure 1.32 allow using in vitro tests to assess the potential bioactivity of a material in a cheap and straightforward method that saves lots of resources to the scientific community. If the results in these tests are negative, there is no need to go forward with cell cultures or in vivo tests. However, if the results are positive, they bring a great amount of information that can be used to implement the materials before using them in cell cultures or animal trials.

1.6.2.3 Bioactivity of Ordered Mesoporous Glasses Compared to Conventional Sol-Gel Glasses

The textural and structural characteristics of ordered mesoporous glasses (MBGs) account for the accelerated bioactive responses compared with traditional bioactive sol-gel glasses with similar compositions (Figure 1.33). In fact, some MBGs give bioactive response after one hour, whereas traditional sol-gel glasses with similar

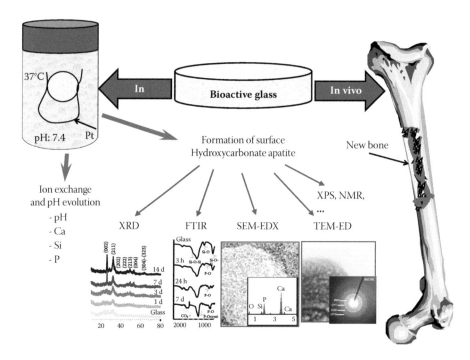

FIGURE 1.32 Different characterization techniques employed to evaluate the bioactive response on the surface of the tested bioceramics after being soaked in physiological solutions, which include analysis of the soaking solution in terms of pH and Ca, Si, and P concentration; X-ray diffraction patterns; Fourier transform infrared spectroscopy; scanning electron microscopy; and transmission electron microscopy.

composition need three days to offer bioactivity. These results can be understood considering different parameters such as specific surface area, pore volume, and diameter, which are more than twice as large as those of conventional sol-gel glasses. In fact, MBGs exhibit the fastest bioactive response ever reported for bioactive materials (Salinas and Vallet-Regí 2007).

In vitro bioactivity assays of MBGs with different calcium content (37%, 20%, and 10% in mol of CaO) were performed (Lopez-Noriega et al. 2006), and some of the results are displayed in Figure 1.34. MBGs are highly bioactive, as indicated by a faster and more intense apatite-like phase formation upon soaking in SBF, compared with traditional sol-gel glasses with similar composition (Arcos, Izquierdo-Barba, and Vallet-Regí 2009).

As already mentioned for traditional sol-gel glasses (Section 1.4.1.2), the main factors that contribute to the crystallization of an apatite phase on the surface are the CaO content and textural properties. A higher amount of CaO leads to lower network connectivity, thus improving the sol-gel glass reactivity. MBGs also show this trend concerning the formation of an amorphous calcium phosphate (ACP) surface layer upon soaking in SBF; i.e., the higher amount of Ca^{2+} released from MBGs containing the highest CaO contents (Si75m and Si58m) leads to the fast formation of a thicker ACP layer compared with Si85m. However, MBGs show a different trend regarding apatite

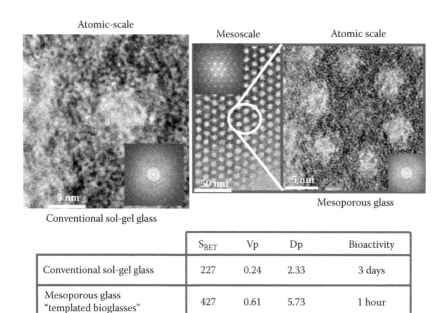

	S$_{BET}$	Vp	Dp	Bioactivity
Conventional sol-gel glass	227	0.24	2.33	3 days
Mesoporous glass "templated bioglasses"	427	0.61	5.73	1 hour

FIGURE 1.33 High-resolution transmission electron microscopy (HRTEM) images and Fourier transform (FT) corresponding to an MBG compared with a conventional sol-gel glass in the SiO_2-CaO-P_2O_5 system in atomic scale and mesoscale. The amorphous pore walls constitute an ordered mesoporous arrangement of cavities in different crystalline systems as the atoms are distributed in an inorganic solid. The table displayed as an inset shows the textural properties (surface area [S$_{BET}$], pore volume [Vp], and pore diameter [Dp]) of both materials with similar composition. The times needed to achieve positive bioactive response are also shown.

crystallization; i.e. the earliest apatite crystallization is observed in MBG with the lowest calcium content (10% CaO). Crystalline hydroxycarbonate apatite is detected by FTIR after 4 hours in Si85m, whereas Si75m and Si58m show the first traces of crystalline calcium phosphate after 24 and 8 hours, respectively. SEM images show characteristic needle-shaped crystalline aggregates of apatite phase after 16 hours in Si85m and after three days in Si75m and Si58m (some traces) (Figure 1.34).

The fast bioactive response of Si85m MBG can be attributed to its higher surface area (427 m^2/g) compared to that of Si75m (195 m^2/g) and Si58m (393 m^2/g). In addition, the influence of pore structure has been postulated as a factor that influences the bioactive behavior of MBGs (Lopez-Noriega et al. 2006). In this sense, Si85m shows a 3D bicontinuous cubic structure that promotes a high diffusion transport mechanism compared with the 2D hexagonal structures of Si75m and Si58m and could favor the accelerated bioactivity behavior of the former. However, when comparing the bioactive behavior of these MBGs, three parameters (composition, textural properties, and structure) are varying together, which makes it difficult to establish the governing factor. To clarify this issue, bioactivity tests were carried out with two MBGs with the same composition, 85SiO$_2$-10CaO-5P$_2$O$_5$, but exhibiting 3D cubic and 2D hexagonal structures, which were synthetized by changing the temperature solvent

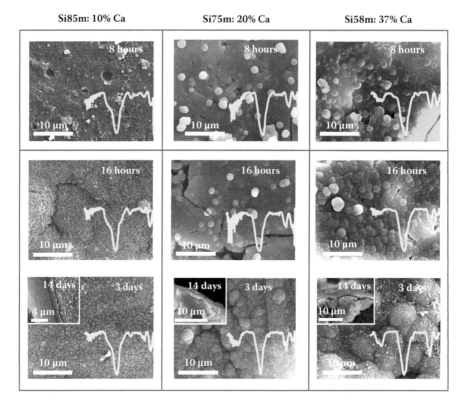

FIGURE 1.34 In vitro bioactivity study of MBGs as function of calcium content (Si85m: $85SiO_2$-$10CaO$-$5P_2O_5$, Si75m: $75SiO_2$-$20CaO$-$5P_2O_5$ and Si58m: $58SiO_2$-$37CaO$-$5P_2O_5$). FTIR spectra after different times in SBF are also displayed. *Source*: Lopez-Noriega et al. (2006).

evaporation (García et al. 2009). Results indicated that both MBGs had identical bioactive behaviors, revealing that differences in mesopore ordering are not decisive for the bioactive behavior of MBGs when composition and texture are equivalents.

An in-depth study by high-resolution TEM has shown again notable differences between Si85m and Si58m BMGs, showing different composition, structural, and textural properties (Izquierdo-Barba et al. 2008). HRTEM study of the MBG with the lowest calcium content (Si85m, 10%CaO) and the highest textural properties shows the formation of nanocrystalline apatite after only one hour in SBF (Figure 1.35.). The 3D bicontinuous cubic structure of this system not only provides high surface area and pore volume but also allows easier ionic exchange with the surrounding medium by increasing the mass transport and diffusion processes. This material exhibits the fastest apatite kinetic formation observed so far for an in vitro biomimetic process.

MBG with the higher calcium content (Si58m, 37%CaO) and 2D hexagonal structure indicated that this is the first known bioceramic that exhibits a sequential transition, namely, amorphous calcium phosphate (ACP)-octacalcium phosphate (OCP)-calcium-deficient carbonate hydroxyapatite (CDHA) as the

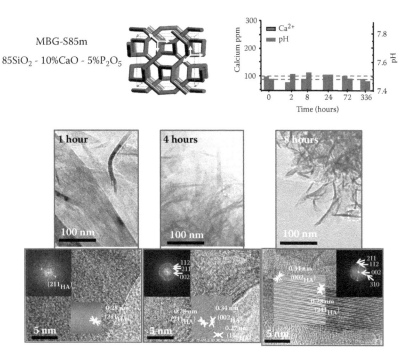

FIGURE 1.35 HRTEM study corresponding to the bioactivity study of MBGS with low calcium content and 3D cubic structure. The inset is a graph showing the values of Ca^{2+} concentration and pH of SBF with the soaking time.

biomineralization-governing mechanism in SBF, similar to the in vivo biomineralization process (Izquierdo-Barba et al. 2008). Usually, all bioceramics prepared to date developed a CHDA phase through the direct crystallization of previously precipitated ACP (Vallet-Regí, Ragel, and Salinas 2003; Vallet-Regí 2001; Hench 2006) without previous formation of the OCP phase, which is formed in the natural bone biomineralization process (Brown, Eidelman, and Tomazic 1987; Nancollas and Tomazic 1974). OCP is a metastable phase and will appear only if the pH in the crystallization system is below 7. In this particular case, this MBG exhibits a high calcium content (37% CaO) and high surface area (195 m^2/g), which provide all factors for a very high reactivity. Consequently, a larger ion exchange takes place between Ca^{2+} and H_3O^+, which leads to a local pH value of 6.5 during the first stage, allowing the formation of metastable OCP (Figure 1.36). However, this local acid pH does not occur in the surfaces of conventional glasses, whose surfaces exhibit the same basic media (pH 7.4 or higher) as the surrounding fluid from the beginning of the bioactive process, which can explain why OCP has never been observed.

The biomimetic bone mineralization of Si58m MBG can be followed by HRTEM, as displayed in Figure 1.37. After one hour soaked in SBF, this MBG generates a large amount of newly formed amorphous calcium phosphate with Ca/P ratio of 1.2. This event has been widely observed in many bioactive compositions, and corresponds to step 4 of the bioactive process described by Hench

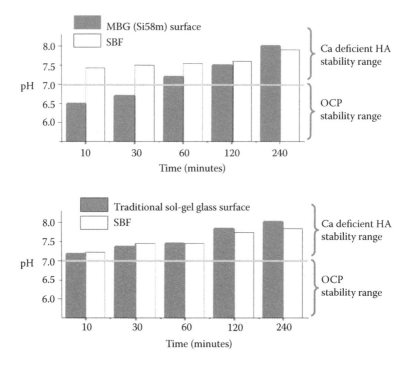

FIGURE 1.36 pH values as a function of time at the bioceramic surface and SBF for Si58m MBG (upper graph) and a traditional sol-gel glass with the same composition (lower graph).

(1991). After four hours of incubation, nanocrystalline oval biphasic nuclei 12 nm in width and 18 nm in length with a Ca/P ratio of 1.3 constituted by OCP with a small fraction of HA were observed. The transformation from oval OCP nuclei to needle-shaped apatite nanocrystals is finally evident on surfaces after eight hours.

The similitude between the in vivo calcium phosphate maturation and the in vitro biomimetism of MBG is tightly related to compositions with relatively high Ca^{2+} content and high open porosity, which forces a reformulation of the bioactivity stages description in terms of the unique physicochemical characteristics exhibited on the MBG surface. Thus the mechanism governing the CDHA layer formation onto the surface of MBG is similar to that proposed by Hench et al. (Hench 1991) for conventional bioactive glasses. However, there are significant differences that lead to an accurate biomimetism in MBG with respect to natural bone. Thus Figure 1.38. shows the proposed bioactivity mechanism of MBGs by comparing with that proposed by Hench for traditional sol-gel glasses, showing the main differences in bioactive behavior (Hench 1991; Izquierdo-Barba et al. 2008; Kaneda et al. 2002; Sakamoto et al. 2004; García et al. 2009; Z. Li, Chen et al. 2007; Kokubo et al. 1990; Kokubo and Takadama 2006).

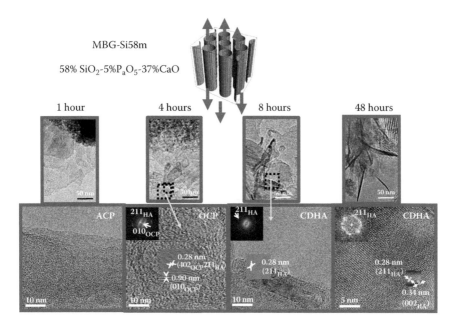

FIGURE 1.37 TEM images of Si58m after being soaked in SBF at different times. After one hour in SBF, a large amount of ACP is observed next to the MBG grain. After four hours in SBF, nanometrical oval nuclei are observed within the ACP matrix. Higher magnification images reveal that these nuclei are nanocrystalline structures constituted by OCP and HA. After eight hours in SBF, needle-shaped apatite crystallizes within the ACP matrix.

The differences in the various stages of the bioactive process can be described as follows:

Stage 1: Rapid exchange of Ca^{2+} with H^+ from the solution (SBF). In the case of MBG, a high and accessible porosity, high surface area, and high material reactivity accelerate the surface processes, which permit a more intense ionic exchange and higher H+ incorporation compared with traditional sol-gel glasses.

Stage 2: Formation of silanol (Si-OH) groups at the glass surface. In the case of MBG, a higher density of Si-OH groups occurs because of the higher surface area and larger H^+ incorporation than traditional sol-gel glasses.

Stage 3: Polycondensation of silanol groups to form a hydrated silica gel. In the case of MBG, a highly protonated silica gel is formed after the condensation of silanol groups, leading to a local acid pH (pH 6.7) on the glass surface. By contrast, this local acid pH does not occur at the surface of conventional sol-gel glasses, whose surface exhibits the same basic media (pH 7.4 or higher) as the surrounding fluid since the beginning of the bioactive process.

Stage 4: Nucleation of an amorphous calcium phosphate (ACP) layer. In the case of MBG, higher ACP precipitation occurs compared with that observed in traditional sol-gel glasses.

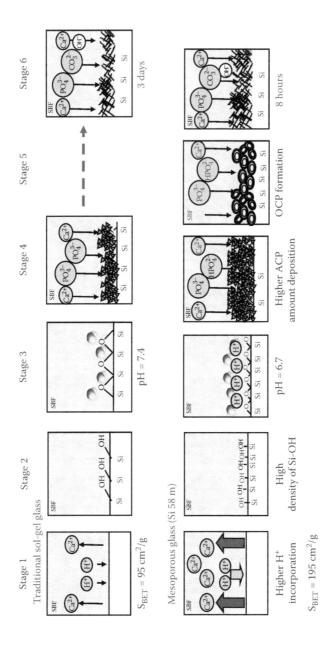

FIGURE 1.38 **(See color insert.)** Bioactive mechanism in SBF of traditional sol-gel glasses (proposed by Hench in 1970) (Hench 1991) compared to bioactive mechanism of MBGs proposed by Izquierdo-Barba et al. (2008).

Stage 5: Crystallization of ACP by incorporation of Ca^{2+} and HPO_4^{2-} ions to form nuclei of octacalcium phosphate (OCP). The acidic surface of the MBG allows the formation of OCP during the first hours. Such a stage does not occur in traditional sol-gel glasses.

Stage 6: In the case of MBG, the maturation of OCP into calcium-deficient hydroxylcarbonate apatite (CDHA) takes place through a dehydration process, which occurs only with small OCP crystallites; the hydrolysis reaction of the PO_4^{3-} and HPO_4^{2-} ions will be the key chemical reaction for OCP to CDHA transformation. In the case of traditional sol-gel glasses, a direct crystallization of previously precipitated ACP (stage 5) occurs by incorporation of OH^- and CO_3^{2-} ions to form HCA. The time period for the CDHA formation is very different in both glasses. In the case of traditional sol-gel glasses, the CDHA formation takes place after three days in SBF, whereas in the MBG it takes only eight hours.

Development of SBF in vitro bioactivity evaluation tests has allowed researchers to establish the mechanism of the bioactive process. It has been also necessary to employ several characterization techniques to provide different views of the same phenomenon.

In 1991 Hench and coworkers proposed for the first time the whole mechanism for bioactive glasses and glass-ceramics (Hench 1991). According to this first theory,

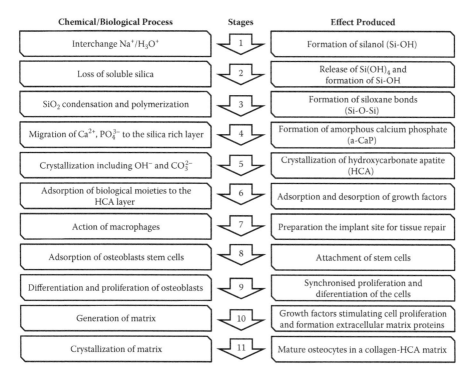

Chemical/Biological Process	Stages	Effect Produced
Interchange Na^+/H_3O^+	1	Formation of silanol (Si-OH)
Loss of soluble silica	2	Release of $Si(OH)_4$ and formation of Si-OH
SiO_2 condensation and polymerization	3	Formation of siloxane bonds (Si-O-Si)
Migration of Ca^{2+}, PO_4^{3-} to the silica rich layer	4	Formation of amorphous calcium phosphate (a-CaP)
Crystallization including OH^- and CO_3^{2-}	5	Crystallization of hydroxycarbonate apatite (HCA)
Adsorption of biological moieties to the HCA layer	6	Adsorption and desorption of growth factors
Action of macrophages	7	Preparation the implant site for tissue repair
Adsorption of osteoblasts stem cells	8	Attachment of stem cells
Differentiation and proliferation of osteoblasts	9	Synchronised proliferation and diferentiation of the cells
Generation of matrix	10	Growth factors stimulating cell proliferation and formation extracellular matrix proteins
Crystallization of matrix	11	Mature osteocytes in a collagen-HCA matrix

FIGURE 1.39 Sequence of the 11 stages taking place on the surface of the glass when in contact with physiological fluids that leads to the formation of new bone.

the bioactive bond between a glass and living bone is formed by a sequence of different reactions that takes place on the surface of the material when in contact with physiological fluids. Figure 1.39 shows the sequence of the 11 stages taking place on the surface of the glass when in contact with physiological fluids, and that leads to the formation of new bone.

The first five stages require a relatively short time, a few days, and can be reproduced in the in vitro bioactivity tests. Sometimes this initial process is so quick that stages 1–3 can be overlapped with stages 4 and 5, and the newly formed layer of hydroxylcarbonate apatite is detected in a few days.

In 1992, Kokubo and coworkers (Li et al. 1992) proposed a slightly different mechanism for bioactivity in CaO-P_2O_5-SiO_2 glasses. These Japanese researchers stated that the most important phenomenon in the formation of the layer was the leaching of the Ca^{2+} ions, which promotes the formation of extra Si-OH groups and saturates the surrounding solution, both factors promoting the hydroxycarbonate apatite. Then once the new layer is formed, the collagen fibers that get to the implant with the initial intention of encapsulating the foreign body find this hydroxycarbonate apatite and help to form the new bone.

The effects on changing the composition of the glasses were observed to have a strong effect on the new layer formation mechanism. Thus when testing glasses without phosphorus, in the system SiO_2-CaO the amorphous calcium phosphate was formed in a few hours, but the hydroxycarbonate apatite formation required seven days. On the other hand, when different amounts of P_2O_5 were used, up to 5% longer times were required for the formation of the amorphous calcium phosphate, but the nanocrystalline hydroxycarbonate apatite was observed after only four days (Salinas, Martin, and Vallet-Regí 2002). When analyzing these different mechanisms with high-resolution transmission electron microscopy (HRTEM) it was found that phosphorus, when present in the composition of the glass, binds to calcium to form Si-doped β-tricalcium phosphate (β-TCP) nanocrystals (Vallet-Regí et al. 2005). In these glasses, the leaching of the calcium ions was impeded, which hampered both the formation of new silanol groups and increased the supersaturation of the solution and therefore longer time was required to form the amorphous calcium phosphate layer. However, once this amorphous layer is formed, the β-TCP nanocrystals act as crystallization nuclei so the crystalline hydroxycarbonate apatite is quickly formed.

In the case of template glasses, the ordered mesoporous structure and high surface area, together with the chemical composition of bioactive glasses, the first five stages were completed in only eight hours after the immersion in SBF (Lopez-Noriega et al. 2006). Analyzing the bioactivity process of these template glasses by HRTEM, it was observed that after this first eight hours, there was octacalcium phosphate (OCP) as the intermediate stage on the surface of these materials (Izquierdo-Barba et al. 2008). This layer evolves then to calcium-deficient carbonate hydroxyapatite, in a similar way to the biomineralization processes that take place in natural bone. Since the discovery of conventional bioactive glasses, much effort has been committed to try to explain the bioactivity of these glasses. Different theories for the bioactive mechanism have been proposed over the years. Figure 1.40 summarizes the different approaches developed so far, which have been described within this section.

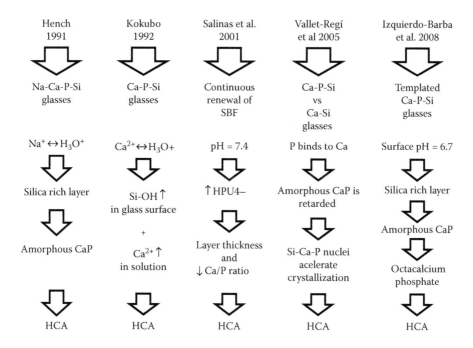

FIGURE 1.40 Different mechanisms proposed for the formation of the new layer of hydroxycarbonated apatite on the surface of bioceramics.

REFERENCES

Alcaide, M., Portolés, A., López-Noriega, D., Acros, M., Vallet-Regí, and M. T. Portolés. 2010. "Interaction of an ordered mesoporous bioactive glass with osteoblasts, fibroblasts and lymphocytes, demonstrating its biocompatibility as a potential bone graft material." *Acta Biomaterialia* 6 (3):892–99.

Arcos, D., M. V. Cabanas, C. V. Ragel, M. Vallet-Regí, and J. San Román. 1997. "Ibuprofen Release from Hydrophilic Ceramic-Polymer Composites." *Biomaterials* 18 (18):1235–42.

Arcos, D., R. P. del Real, and M. Vallet-Regí. 2002. "A Novel Bioactive and Magnetic Biphasic Material." *Biomaterials* 23 (10):2151–58.

———. 2003. "Biphasic Materials for Bone Grafting and Hyperthermia Treatment of Cancer." *Journal of Biomedical Materials Research Part A* 65A (1):71–78.

Arcos, D., D. C. Greenspan, and M. Vallet-Regí. 2002. "Influence of the Stabilization Temperature on Textural and Structural Features and Ion Release in SiO2-CaO-P2O5 Sol-Gel Glasses." *Chemistry of Materials* 14 (4):1515–22.

Arcos, D., I. Izquierdo-Barba, and M. Vallet-Regí. 2009. "Promising Trends of Bioceramics in the Biomaterials Field." *Journal of Materials Science: Materials in Medicine* 20 (2):447–55.

Arias, J. L., M. B. Mayor, J. Pou, Y. Leng, B. Leon, and M. Perez-Amor. 2003. "Micro- and Nano-Testing of Calcium Phosphate Coatings Produced by Pulsed Laser Deposition." *Biomaterials* 24 (20):3403–8.

Avnir, D., T. Coradin, O. Lev, and J. Livage. 2006. "Recent Bio-applications of Sol-Gel Materials." *Journal of Materials Chemistry* 16 (11):1013–30.

Bagshaw, S. A., E. Prouzet, and T. J. Pinnavaia. 1995. "Templating of Mesoporous Molecular-Sieves by Nonionic Polyethylene Oxide Surfactants." *Science* 269 (5228):1242–44.

Balas, F., D. Arcos, J. Perez-Pariente, and M. Vallet-Regí. 2001. "Textural Properties of SiO2 Center Dot CaO Center Dot P2O5 Glasses Prepared by the Sol-Gel Method." *Journal of Materials Research* 16 (5):1345–48.

Beck, J. S., J. C. Vartuli, W. J. Roth, M. E. Leonowicz, C. T. Kresge, K. D. Schmitt, C. T. W. Chu, D. H. Olson, E. W. Sheppard, S. B. McCullen, J. B. Higgins, and J. L. Schlenker. 1992. "A New Family of Mesoporous Molecular-Sieves Prepared with Liquid-Crystal Templates." *Journal of the American Chemical Society* 114 (27):10834–43.

Brink, M. 1997. "The Influence of Alkali and Alkaline Earths on the Working Range for Bioactive Glasses." *Journal of Biomedical Materials Research* 36 (1):109–17.

Brink, M., T. Turunen, R. P. Happonen, and A. Yli-Urpo. 1997. "Compositional Dependence of Bioactivity of Glasses in the System Na2O-K2O-MgO-CaO-B2O3-P2O5-SiO2." *Journal of Biomedical Materials Research* 37 (1):114–21.

Brinker, C. J., Y. F. Lu, A. Sellinger, and H. Y. Fan. 1999. "Evaporation-Induced Self-Assembly: Nanostructures Made Easy." *Advanced Materials* 11 (7):579–85.

Brown, W. E., N. Eidelman, and B. Tomazic. 1987. "Octacalcium Phosphate as a Precursor in Biomineral Formation." *Advances in Dental Research* 1 (2):306–13.

Cabanas, M. V., L. M. Rodriguez-Lorenzo, and M. Vallet-Regí. 2002. "Setting Behavior and In Vitro Bioactivity of Hydroxyapatite/Calcium Sulfate Cements." *Chemistry of Materials* 14 (8):3550–55.

Cabanas, M. V., and M. Vallet-Regí. 2003. "Calcium Phosphate Coatings Deposited by Aerosol Chemical Vapour Deposition." *Journal of Materials Chemistry* 13 (5):1104–7.

Che, S., A. E. Garcia-Bennett, T. Yokoi, K. Sakamoto, H. Kunieda, O. Terasaki, and T. Tatsumi. 2003. "A Novel Anionic Surfactant Templating Route for Synthesizing Mesoporous Silica with Unique Structure." *Nature Materials* 2 (12):801–5.

Chen, Q., F. Miyaji, T. Kokubo, and T. Nakamura. 1999. "Apatite Formation on PDMS-Modified CaO-SiO2-TiO2 Hybrids Prepared by Sol-Gel Process." *Biomaterials* 20 (12):1127–32.

Chen, Q., N. Miyata, T. Kokubo, and T. Nakamura. 2000. "Bioactivity and Mechanical Properties of PDMS-Modified CaO-SiO2-TiO2 Hybrids Prepared by Sol-Gel Process." *Journal of Biomedical Materials Research* 51 (4):605–11.

Chow, L. C. 1991. "Development of Self-Setting Calcium-Phosphate Cements." *Nippon Seramikkusu Kyokai Gakujutsu Ronbunshi (Journal of the Ceramic Society of Japan)* 99 (10):954–64.

Cicuéndez, Mónica, Isabel Izquierdo-Barba, Sandra Sánchez-Salcedo, Mercedes Vila, and María Vallet-Regí. 2012. "Biological Performance of Hydroxyapatite-Biopolymer Foams: In Vitro Cell Response." *Acta Biomaterialia* 8 (2):802–10.

Ciesla, U., and F. Schuth. 1999. "Ordered Mesoporous Materials." *Microporous and Mesoporous Materials* 27 (2–3):131–49.

Coetzee, A. S. 1980. "Regeneration of Bone in the Presence of Calcium-Sulfate." *Archives of Otolaryngology—Head & Neck Surgery* 106 (7):405–9.

Colilla, M., A. J. Salinas, and M. Vallet-Regí. 2006. "Amino-Polysiloxane Hybrid Materials for Bone Reconstruction." *Chemistry of Materials* 18 (24):5676–83.

Constantz, B. R., I. C. Ison, M. T. Fulmer, R. D. Poser, S. T. Smith, M. Vanwagoner, J. Ross, S. A. Goldstein, J. B. Jupiter, and D. I. Rosenthal. 1995. "Skeletal Repair by in-Situ Formation of the Mineral Phase of Bone." *Science* 267 (5205):1796–99.

Coradin, T., M. Boissiere, and J. Livage. 2006. "Sol-Gel Chemistry in Medicinal Science." *Current Medicinal Chemistry* 13 (1):99–108.

Daculsi, G. 1998. "Biphasic Calcium Phosphate Concept Applied to Artificial Bone, Implant Coating and Injectable Bone Substitute." *Biomaterials* 19 (16):1473–78.

Day, R. M., and A. R. Boccaccini. 2005. "Effect of Particulate Bioactive Glasses on Human Macrophages and Monocytes In Vitro." *Journal of Biomedical Materials Research Part A* 73A (1):73–79.

Degroot, K. 1980. "Bioceramics Consisting of Calcium-Phosphate Salts." *Biomaterials* 1 (1):47–50.

Deguchi, K., K. Tsuru, T. Hayashi, M. Takaishi, M. Nagahara, S. Nagotani, Y. Sehara, G. Jin, H. Z. Zhang, S. Hayakawa, M. Shoji, M. Miyazaki, A. Osaka, N. H. Huh, and K. Abe. 2006. "Implantation of a New Porous Gelatin-Siloxane Hybrid into a Brain Lesion as a Potential Scaffold for Tissue Regeneration." *Journal of Cerebral Blood Flow and Metabolism* 26 (10):1263–73.

del Real, R. P., E. Ooms, J. G. C. Wolke, M. Vallet-Regí, and J. A. Jansen. 2003. "In Vivo Bone Response to Porous Calcium Phosphate Cement." *Journal of Biomedical Materials Research Part A* 65A (1):30–36.

del Real, R. P., S. Padilla, and M. Vallet-Regí. 2000. "Gentamicin Release from Hydroxyapatite/poly(ethyl methacrylate)/poly(methyl methacrylate) Composites." *Journal of Biomedical Materials Research* 52 (1):1–7.

Doi, Y., T. Shibutani, Y. Moriwaki, T. Kajimoto, and Y. Iwayama. 1998. "Sintered Carbonate Apatites as Bioresorbable Bone Substitutes." *Journal of Biomedical Materials Research* 39 (4):603–10.

Dorozhkin, S. V., and M. Epple. 2002. "Biological and Medical Significance of Calcium Phosphates." *Angewandte Chemie—International Edition* 41 (17):3130–46.

Du, R. L., and J. Chang. 2004. "Preparation and Characterization of Zn and Mg Doped Bioactive Glasses." *Journal of Inorganic Materials* 19 (6):1353–58.

Du, R. L., J. Chang, S. Y. Ni, W. Y. Zhai, and J. Y. Wang. 2006. "Characterization and In Vitro Bioactivity of Zinc-Containing Bioactive Glass and Glass-Ceramics." *Journal of Biomaterials Applications* 20 (4):341–60.

Ebisawa, Y., F. Miyaji, T. Kokubo, K. Ohura, and T. Nakamura. 1997a. "Bioactivity of Ferrimagnetic Glass-Ceramics in the System FeO-Fe2O3-CaO-SiO2." *Biomaterials* 18 (19):1277–84.

———. 1997b. "Surface Reaction of Bioactive and Ferrimagnetic Glass-Ceramics in the System FeO-Fe2O3-CaO-SiO2." *Journal of the Ceramic Society of Japan* 105 (11):947–51.

Ebisawa, Y., Y. Sugimoto, T. Hayashi, T. Kokubo, K. Ohura, and T. Yamamuro. 1991. "Crystallization of (FEO, FE2O3)-CAO-SIO2 Glasses and Magnetic Properties of Their Crystallized Products." *Nippon Seramikkusu Kyokai Gakujutsu Ronbunshi (Journal of the Ceramic Society of Japan)* 99 (1):7–13.

Elliott, J. C. 1994. *Structure and Chemistry of the Apatites and Other Calcium Orthophosphates.* Amsterdam: Elsevier.

Fan, J., C. Z. Yu, T. Gao, J. Lei, B. Z. Tian, L. M. Wang, Q. Luo, B. Tu, W. Z. Zhou, and D. Y. Zhao. 2003. "Cubic Mesoporous Silica with Large Controllable Entrance Sizes and Advanced Adsorption Properties." *Angewandte Chemie-International Edition* 42 (27):3146–50.

Firouzi, A., F. Atef, A. G. Oertli, G. D. Stucky, and B. F. Chmelka. 1997. "Alkaline Lyotropic Silicate-Surfactant Liquid Crystals." *Journal of the American Chemical Society* 119 (15):3596–3610.

García, A., M. Cicuéndez, I. Izquierdo-Barba, D. Arcos, and M. Vallet-Regí. 2009. "Essential Role of Calcium Phosphate Heterogeneities in 2D-Hexagonal and 3D-Cubic SiO2-CaO-P2O5 Mesoporous Bioactive Glasses." *Chemistry of Materials* 21 (22):5474–84.

García, A., I. Izquierdo-Barba, M. Colilla, C. L. de Laorden, and M. Vallet-Regí. 2011. "Preparation of 3-D Scaffolds in the SiO(2)-P(2)O(5) System with Tailored Hierarchical Meso-Macroporosity." *Acta Biomaterialia* 7 (3):1265–73.

Gauthier, O., E. Goyenvalle, J. M. Bouler, J. Guicheux, P. Pilet, P. Weiss, and G. Daculsi. 2001. "Macroporous Biphasic Calcium Phosphate Ceramics versus Injectable Bone Substitute: A Comparative Study 3 and 8 Weeks after Implantation in Rabbit Bone." *Journal of Materials Science-Materials in Medicine* 12 (5) 385–90.

Gil-Albarova, J., R. Garrido-Lahiguera, A. J. Salinas, J. Roman, A. Bueno-Lozano, R. Gil-Albarova, and M. Vallet-Regí. 2004. "The In Vivo Performance of a Sol-Gel Glass and a Glass-Ceramic in the Treatment of Limited Bone Defects." *Biomaterials* 25 (19):4639–45.

Gil-Albarova, J., A. J. Salinas, A. L. Bueno-Lozano, J. Roman, N. Aldini-Nicolo, A. García-Barea, G. Giavaresi, M. Fini, R. Giardino, and M. Vallet-Regí. 2005. "The In Vivo Behaviour of a Sol-Gel Glass and a Glass-Ceramic during Critical Diaphyseal Bone Defects Healing." *Biomaterials* 26 (21):4374–82.

Ginebra, M. P., F. C. M. Driessens, and J. A. Planell. 2004. "Effect of the Particle Size on the Micro and Nanostructural Features of a Calcium Phosphate Cement: A Kinetic Analysis." *Biomaterials* 25 (17):3453–62.

Gonzalez, B., M. Colilla, and M. Vallet-Regí. 2008. "Time-Delayed Release of Bioencapsulates: A Novel Controlled Delivery Concept for Bone Implant Technologies." *Chemistry of Materials* 20 (15):4826–34.

Granado, S., V. Ragel, V. Cabanas, J. SanRoman, and M. Vallet-Regí. 1997. "Influence of Alpha-Al2O3 Morphology and Particle Size on Drug Release from Ceramic/Polymer Composites." *Journal of Materials Chemistry* 7 (8):1581–85.

Greenspan, D. C., and l. L. Hench. 1976. "Chemical and Mechanical-Behavior of Bioglass-Coated Alumina." *Journal of Biomedical Materials Research* 10 (4):503–9.

Grimandi, G., P. Weiss, F. Millot, and G. Daculsi. 1998. "In Vitro Evaluation of a New Injectable Calcium Phosphate Material." *Journal of Biomedical Materials Research* 39 (4):660–66.

Gunawidjaja, Philips N., Renny Mathew, Andy Y. H. Lo, Isabel Izquierdo-Barba, Ana García, Daniel Arcos, María Vallet-Regí, and Mattias Edén. 2012. "Local Structures of Mesoporous Bioactive Glasses and Their Surface Alterations In Vitro: Inferences from Solid-State Nuclear Magnetic Resonance." *Philosophical Transactions of the Royal Society A: Mathematical, Physical and Engineering Sciences* 370 (1963):1376–99.

Hench, L. L. 1991. "Bioceramics—from Concept to Clinic." *Journal of the American Ceramic Society* 74 (7):1487–1510.

Hench, L. L., R. J. Splinter, W. C. Allen, and T. K. Greenlee. 1971. "Bonding Mechanisms at the Interface of Ceramic Prosthetic Materials." *Journal of Biomedical Materials Research* 5 (6):117–41.

Hench, L. L., and J. K. West. 1990. "The Sol-Gel Process." *Chemical Reviews* 90 (1):33–72.

Hench, L. L., and J. Wilson. 1984. "Surface-Active Biomaterials." *Science* 226 (4675):630–36.

Hench, Larry. 2006. "The Story of Bioglass®." *Journal of Materials Science: Materials in Medicine* 17 (11):967–78.

Hijon, N., M. V. Cabanas, I. Izquierdo-Barba, M. A. García, and M. Vallet-Regí. 2006. "Nanocrystalline Bioactive Apatite Coatings." *Solid State Sciences* 8 (6):685–91.

Hijon, N., M. V. Cabanas, I. Izquierdo-Barba, and M. Vallet-Regí. 2004. "Bioactive Carbonate-Hydroxyapatite Coatings Deposited onto Ti6Al4V Substrate." *Chemistry of Materials* 16 (8):1451–55.

Hijon, N., M. Manzano, A. J. Salinas, and M. Vallet-Regí. 2005. "Bioactive CaO-SiO2-PDMS Coatings on Ti6Al4V Substrates." *Chemistry of Materials* 17 (6):1591–96.

Hoffmann, F., M. Cornelius, J. Morell, and M. Froba. 2006. "Silica-Based Mesoporous Organic-Inorganic Hybrid Materials." *Angewandte Chemie-International Edition* 45 (20):3216–51.

Huang, Y. T., M. Imura, Y. Nemoto, C. H. Cheng, and Y. Yamauchi. 2011. "Block-Copolymer-Assisted Synthesis of Hydroxyapatite Nanoparticles with High Surface Area and Uniform Size." *Science and Technology of Advanced Materials* 12 (4). doi:10.1088/1468-6996/12/4/045005

Ikawa, N., H. Hori, T. Kimura, Y. Oumi, and T. Sano. 2008. "Templating Route for Mesostructured Calcium Phosphates with Carboxylic Acid- and Amine-Type Surfactants." *Langmuir* 24 (22):13113–20.

———. 2011. "Unique Surface Property of Surfactant-Assisted Mesoporous Calcium Phosphate." *Microporous and Mesoporous Materials* 141 (1–3):56–60.

Ikawa, N., Y. Oumi, T. Kimura, T. Ikeda, and T. Sano. 2008. "Synthesis of Lamellar Mesostructured Calcium Phosphates Using N-alkylamines as Structure-Directing Agents in Alcohol/Water Mixed Solvent Systems." *Journal of Materials Science* 43 (12):4198–4207.

Ikenaga, M., K. Ohura, T. Yamamuro, Y. Kotoura, M. Oka, and T. Kokubo. 1993. "Localized Hyperthermic Treatment of Experimental Bone Tumors with Ferromagnetic Ceramics." *Journal of Orthopaedic Research* 11 (6):849–55.

Inagaki, S., A. Koiwai, N. Suzuki, Y. Fukushima, and K. Kuroda. 1996. "Syntheses of Highly Ordered Mesoporous Materials, FSM-16, Derived from Kanemite." *Bulletin of the Chemical Society of Japan* 69 (5):1449–57.

Izquierdo-Barba, I., D. Arcos, Y. Sakamoto, O. Terasaki, A. Lopez-Noriega, and M. Vallet-Regí. 2008. "High-Performance Mesoporous Bioceramics Mimicking Bone Mineralization." *Chemistry of Materials* 20 (9):3191–98.

Izquierdo-Barba, I., A. Asenjo, L. Esquivias, and M. Vallet-Regí. 2003. "SiO2-CaO Vitreous Films Deposited onto Ti6Al4V Substrates." *European Journal of Inorganic Chemistry* (8):1608–13.

Izquierdo-Barba, I., F. Conde, N. Olmo, M. A. Lizarbe, M. A. García, and M. Vallet-Regí. 2006. "Vitreous SiO2-CaO Coatings on Ti6Al4V Alloys: Reactivity in Simulated Body Fluid versus Osteoblast Cell Culture." *Acta Biomaterialia* 2 (4):445–55.

Izquierdo-Barba, I., A. J. Salinas, and M. Vallet-Regí. 2000. "Effect of the Continuous Solution Exchange on the In Vitro Reactivity of a CaO-SiO2 Sol-Gel Glass." *Journal of Biomedical Materials Research* 51 (2):191–99.

Izquierdo-Barba, I., M. Vallet-Regí, J. M. Rojo, E. Blanco, and L. Esquivias. 2003. "The Role of Precursor Concentration on the Characteristics of SiO2-CaO Films." *Journal of Sol-Gel Science and Technology* 26 (1–3):1179–82.

Jha, L. J., S. M. Best, J. C. Knowles, I. Rehman, J. D. Santos, and W. Bonfield. 1997. "Preparation and Characterization of Fluoride-Substituted Apatites." *Journal of Materials Science: Materials in Medicine* 8 (4):185–91.

Kaneda, M., T. Tsubakiyama, A. Carlsson, Y. Sakamoto, T. Ohsuna, O. Terasaki, S. H. Joo, and R. Ryoo. 2002. "Structural Study of Mesoporous MCM-48 and Carbon Networks Synthesized in the Spaces of MCM-48 by Electron Crystallography." *Journal of Physical Chemistry B* 106 (6):1256–66.

Kannan, S., A. Balamurugan, and S. Rajeswari. 2002. "Development of Calcium Phosphate Coatings on Type 316L SS and Their In Vitro Response." *Trends in Biomaterials and Artificial Organs* 16 (1):8–11.

Kivrak, N., and A. C. Tas. 1998. "Synthesis of Calcium Hydroxyapatite-Tricalcium Phosphate (HA-TCP) Composite Bioceramic Powders and Their Sintering Behavior." *Journal of the American Ceramic Society* 81 (9):2245–52.

Koch, B., J. G. C. Wolke, and K. de Groot. 1990. "X-Ray-Diffraction Studies on Plasma-Sprayed Calcium Phosphate-Coated Implants." *Journal of Biomedical Materials Research* 24 (6):655–67.

Kokubo, T. 1991. "Bioactive Glass-Ceramics—Properties and Applications." *Biomaterials* 12 (2):155–63.

Kokubo, T., H. M. Kim, and M. Kawashita. 2003. "Novel Bioactive Materials with Different Mechanical Properties." *Biomaterials* 24 (13):2161–75.

Kokubo, T., H. Kushitani, S. Sakka, T. Kitsugi, and T. Yamamuro. 1990. "Solutions Able to Reproduce in Vivo Surface-Structure Changes in Bioactive Glass-Ceramic A-W3." *Journal of Biomedical Materials Research* 24 (6):721–34.

Kokubo, T., and H. Takadama. 2006. "How Useful Is SBF in Predicting In Vivo Bone Bioactivity?" *Biomaterials* 27 (15):2907–15.

Kresge, C. T., M. E. Leonowicz, W. J. Roth, J. C. Vartuli, and J. S. Beck. 1992. "Ordered Mesoporous Molecular Sieves Synthesized by a Liquid-Crystal Template Mechanism." *Nature* 359 (6397):710–12.

León, B., and J. Jansen. 2009. *Thin Calcium Phosphate Coatings for Medical Implants.* New York: Springer.

Leonova, E., I. Izquierdo-Barba, D. Arcos, A. Lopez-Noriega, N. Hedin, M. Vallet-Regí, and M. Eden. 2008. "Multinuclear Solid-State NMR Studies of Ordered Mesoporous Bioactive Glasses." *Journal of Physical Chemistry C* 112 (14):5552–62.

Li, P., C. Ohtsuki, T. Kokubo, K. Nakanishi, N. Soga, T. Nakamura, and T. Yamamuro. 1992. "Apatite Formation Induced by Silica Gel in a Simulated Body Fluid." *Journal of the American Ceramic Society* 75 (8):2094–2097.

Li, R., A. E. Clark, and L. L. Hench. 1991. "Hierarchically Porous Bioactive Glass Scaffolds Synthesized with a PUF and P123 Cotemplated Approach." *Chemistry of Materials* 19 (17):4322–26.

Li, Xia, Jianlin Shi, Xiaoping Dong, Lingxia Zhang, and Hongyu Zeng. 2008. "A Mesoporous Bioactive Glass/Polycaprolactone Composite Scaffold and Its Bioactivity Behavior." *Journal of Biomedical Materials Research Part A* 84A (1):84–91.

Li, Zheng, Dehong Chen, Bo Tu, and Dongyuan Zhao. 2007. "Synthesis and Phase Behaviors of Bicontinuous Cubic Mesoporous Silica from Triblock Copolymer Mixed Anionic Surfactant." *Microporous and Mesoporous Materials* 105 (1–2):34–40.

Livingston, T., P. Ducheyne, and J. Garino. 2002. "In Vivo Evaluation of a Bioactive Scaffold for Bone Tissue Engineering." *Journal of Biomedical Materials Research* 62 (1):1–13.

Lopez-Noriega, A., D. Arcos, I. Izquierdo-Barba, Y. Sakamoto, O. Terasaki, and M. Vallet-Regí. 2006. "Ordered Mesoporous Bioactive Glasses for Bone Tissue Regeneration." *Chemistry of Materials* 18 (13):3137–44.

Manzano, M., D. Arcos, M. Redríguez-Delgado, E. Ruiz-Hernández, F. J. Gil, and M. Vallet-Regí. 2006. "Bioactive Star Gels." *Chemistry of Materials* 18:5696–5703.

Manzano, M., A. J. Salinas, and M. Vallet-Regí. 2006. "P-containing Ormosils for Bone Reconstruction." *Progress in Solid State Chemistry* 34 (2–4):267–77.

Marques, Paap, M. C. F. Magalhaes, R. N. Correia, A. I. Martin, A. J. Salinas, and M. Vallet-Regí. 2004. "Ceramics In Vitro Mineralisation Protocols: A Supersaturation Problem." *Bioceramics* 16:143–46. doi:10.4028/www.scientific.net/KEM.254-256.143

Martin, A. I., A. J. Salinas, and M. Vallet-Regí. 2005. "Bioactive and Degradable Organic-Inorganic Hybrids." *Journal of the European Ceramic Society* 25 (16):3533–38.

Martinez, A., I. Izquierdo-Barba, and M. Vallet-Regí. 2000. "Bioactivity of a CaO-SiO2 Binary Glasses System." *Chemistry of Materials* 12 (10):3080–88.

Meng, Y., D. Gu, F. Q. Zhang, Y. F. Shi, H. F. Yang, Z. Li, C. Z. Yu, B. Tu, and D. Y. Zhao. 2005. "Ordered Mesoporous Polymers and Homologous Carbon Frameworks: Amphiphilic Surfactant Templating and Direct Transformation." *Angewandte Chemie—International Edition* 44 (43):7053–59.

Meseguer-Olmo, L., M. J. Ros-Nicolas, M. Clavel-Sainz, V. Vicente-Ortega, M. Alcaraz-Banos, A. Lax-Perez, D. Arcos, C. V. Ragel, and M. Vallet-Regí. 2002. "Biocompatibility and In Vivo Gentamicin Release from Bioactive Sol-Gel Glass Implants." *Journal of Biomedical Materials Research* 61 (3):458–65.

Miyamoto, Y., K. Ishikawa, M. Takechi, T. Toh, T. Yuasa, M. Nagayama, and K. Suzuki. 1998. "Basic Properties of Calcium Phosphate Cement Containing Atelocollagen in Its Liquid or Powder Phases." *Biomaterials* 19 (7–9):707–15.

Miyata, N., K. Fuke, Q. Chen, M. Kawashita, T. Kokubo, and T. Nakamura. 2002. "Apatite-Forming Ability and Mechanical Properties of PTMO-modified CaO-SiO2 Hybrids Prepared by Sol-Gel Processing: Effect of CaO and PTMO Contents." *Biomaterials* 23 (14):3033–40.

———. 2004. "Apatite-Forming Ability and Mechanical Properties of PTMO-Modified CaO-SiO2-TiO2 Hybrids Derived from Sol-Gel Processing." *Biomaterials* 25 (1):1–7.

Nancollas, G. H., and B. Tomazic. 1974. "Growth of Calcium Phosphate on Hydroxyapatite Crystals: Effect of Supersaturation and Ionic Medium." *The Journal of Physical Chemistry* 78 (22):2218–25.

Narasaraju, T. S. B., and D. E. Phebe. 1996. "Some Physico-Chemical Aspects of Hydroxylapatite." *Journal of Materials Science* 31 (1):1–21.

Ng, S. X., J. Guo, J. Ma, and S. C. J. Loo. 2010. "Synthesis of High Surface Area Mesostructured Calcium Phosphate Particles." *Acta Biomaterialia* 6 (9):3772–81.

Nieto, A., S. Areva, T. Wilson, R. Viitala, and M. Vallet-Regí. 2009. "Cell Viability in a Wet Silica Gel. *Acta Biomaterialia* 5 (9):3478–87.

Nilsson, M., E. Fernandez, S. Sarda, L. Lidgren, and J. A. Planell. 2002. "Characterization of a Novel Calcium Phosphate/Sulphate Bone Cement." *Journal of Biomedical Materials Research* 61 (4):600–607.

Ohura, K., M. Ikenaga, T. Nakamura, T. Yamamuro, Y. Ebisawa, T. Kokubo, Y. Kotoura, and M. Oka. 1991. "A Heat-Generating Bioactive Glass Ceramic for Hyperthermia." *Journal of Applied Biomaterials* 2 (3):153–59.

Oki, A., B. Parveen, S. Hossain, S. Adeniji, and H. Donahue. 2004. "Preparation and In Vitro Bioactivity of Zinc Containing Sol-Gel-Derived Bioglass Materials." *Journal of Biomedical Materials Research Part A* 69A (2):216–21.

Olmo, N., A. I. Martin, A. J. Salinas, J. Turnay, M. Vallet-Regi, and M. A. Lizarbe. 2003. "Bioactive Sol-Gel Glasses with and without a Hydroxycarbonate Apatite Layer as Substrates for Osteoblast Cell Adhesion and Proliferation." *Biomaterials* 24 (20):3383–93.

Oonishi, H., L. L. Hench, J. Wilson, F. Sugihara, E. Tsuji, M. Matsuura, S. Kin, T. Yamamoto, and S. Mizokawa. 2000. "Quantitative Comparison of Bone Growth Behavior in Granules of Bioglass®, A-W Glass-Ceramic, and Hydroxyapatite." *Journal of Biomedical Materials Research* 51 (1):37–46.

Ostomel, T. A., Q. H. Shi, C. K. Tsung, H. J. Liang, and G. D. Stucky. 2006. "Spherical Bioactive Glass with Enhanced Rates of Hydroxyapatite Deposition and Hemostatic Activity." *Small* 2 (11):1261–65.

Otsuka, M., Y. Matsuda, Y. Suwa, J. L. Fox, and W. I. Higuchi. 1995. "A Novel Skeletal Drug-Delivery System Using Self-Setting Calcium-Phosphate Cement. 6: Effect of Particle-Size of Metastable Calcium Phosphates on Mechanical Strength of a Novel Self-Setting Bioactive Calcium-Phosphate Cement." *Journal of Biomedical Materials Research* 29 (1):25–32.

Ozin, G. A., N. Varaksa, N. Coombs, J. E. Davies, D. D. Perovic, and M. Ziliox. 1997. "Bone Mimetics: A Composite of Hydroxyapatite and Calcium Dodecylphosphate Lamellar Phase." *Journal of Materials Chemistry* 7 (8):1601–7.

Padilla, S., R. P. del Real, and M. Vallet-Regí. 2002. "In Vitro Release of Gentamicin from OHAp/PEMA/PMMA Samples." *Journal of Controlled Release* 83 (3):343–52.

Padilla, S., J. Roman, S. Sanchez-Salcedo, and M. Vallet-Regi. 2006. "Hydroxyapatite/SiO2-CaO-P2O5 Glass Materials: In Vitro Bioactivity and Biocompatibility." *Acta Biomaterialia* 2 (3):331–42.

Padilla, S., S. Sanchez-Salcedo, and M. Vallet-Regí. 2005. "Bioactive and Biocompatible Pieces of HA/sol-gel Glass Mixtures Obtained by the Gel-Casting Method." *Journal of Biomedical Materials Research Part A* 75A (1):63–72.

Park, Y. S., K. Y. Yi, I. S. Lee, C. H. Han, and Y. C. Jung. 2005. "The Effects of Ion Beam-Assisted Deposition of Hydroxyapatite on the Grit-Blasted Surface of Endosseous Implants in Rabbit Tibiae." *International Journal of Oral & Maxillofacial Implants* 20 (1):31–38.

Pechini, M. P. 1967. "Method for Preparing Lead and Alkaline Earth Titanates and Niobates and Coating Method Using the Same to Form Capacitor." U.S. Patent 3330697.

Peltola, T., M. Jokinen, H. Rahiala, E. Levanen, J. B. Rosenholm, I. Kangasniemi, and A. Yli-Urpo. 1999. "Calcium Phosphate Formation on Porous Sol-Gel-Derived SiO2 and CaO-P2O5-SiO2 Substrates In Vitro." *Journal of Biomedical Materials Research* 44 (1):12–21.

Pena, J., and M. Vallet-Regí. 2003. "Hydroxyapatite, Tricalcium Phosphate and Biphasic Materials Prepared by a Liquid Mix Technique." *Journal of the European Ceramic Society* 23 (10):1687–96.

Pereira, A. P. V., W. L. Vasconcelos, and R. L. Orefice. 2000. "Novel Multicomponent Silicate-Poly(Vinyl Alcohol) Hybrids with Controlled Reactivity." *Journal of Non-Crystalline Solids* 273 (1–3):180–85.

Pereira, M. M., A. E. Clark, and L. L. Hench. 1994. "Calcium-Phosphate Formation on Sol-Gel-Derived Bioactive Glasses In-Vitro." *Journal of Biomedical Materials Research* 28 (6):693–98.

Perez-Pariente, J., F. Balas, J. Roman, A. J. Salinas, and M. Vallet-Regí. 1999. "Influence of Composition and Surface Characteristics on the In Vitro Bioactivity of SiO2-CaO-P2O5-MgO Sol-Gel Glasses." *Journal of Biomedical Materials Research* 47 (2):170–75.

Perez-Pariente, J., F. Balas, and M. Vallet-Regí. 2000. "Surface and Chemical Study of SiO2 Center Dot P2O5 Center Dot CaO Center Dot(MgO) Bioactive Glasses." *Chemistry of Materials* 12 (3):750–55.

Pietrzak, W. S., and R. Ronk. 2000. "Calcium Sulfate Bone Void Filler: A Review and a Look Ahead." *Journal of Craniofacial Surgery* 11 (4):27–33.

Prelot, B., and T. Zemb. 2005. "Calcium Phosphate Precipitation in Catanionic Templates." *Materials Science & Engineering C-Biomimetic and Supramolecular Systems* 25 (5–8):553–59.

Ragel, C. V., M. Vallet-Regí, and L. M. Rodriguez-Lorenzo. 2002. "Preparation and In Vitro Bioactivity of Hydroxyapatite/Solgel Glass Biphasic Material." *Biomaterials* 23 (8):1865–72.

Ramila, A., S. Padilla, B. Munoz, and M. Vallet-Regí. 2002. "A New Hydroxyapatite/Glass Biphasic Material: In Vitro Bioactivity." *Chemistry of Materials* 14 (6):2439–43.

Ravikovitch, P. I., and A. V. Neimark. 2002. "Density Functional Theory of Adsorption in Spherical Cavities and Pore Size Characterization of Templated Nanoporous Silicas with Cubic and Three-Dimensional hexagonal structures." *Langmuir* 18 (5):1550–60.

———. 2002. "Experimental Confirmation of Different Mechanisms of Evaporation from Ink-Bottle Type Pores: Equilibrium, Pore Blocking, and Cavitation." *Langmuir* 18 (25):9830–37.

Ruiz-Hernández, E., A. Lopez-Noriega, D. Arcos, I. Izquierdo-Barba, O. Terasaki, and M. Vallet-Regí. 2007. "Aerosol-Assisted Synthesis of Magnetic Mesoporous Silica Spheres for Drug Targeting." *Chemistry of Materials* 19 (14):3455–63.

Ruiz-Hernández, E., A. Lopez-Noriega, D. Arcos, and M. Vallet-Regí. 2008. "Mesoporous Magnetic Microspheres for Drug Targeting." *Solid State Sciences* 10 (4):421–26.

Ruiz-Hernández, E., M. C. Serrano, D. Arcos, and M. Vallet-Regí. 2006. "Glass-Glass Ceramic Thermoseeds for Hyperthermic Treatment of Bone Tumors." *Journal of Biomedical Materials Research Part A* 79A (3):533–43.

Ryoo, R., J. M. Kim, C. H. Ko, and C. H. Shin. 1996. "Disordered Molecular Sieve with Branched Mesoporous Channel Network." *Journal of Physical Chemistry* 100 (45):17718–21.

Sadasivan, S., D. Khushalani, and S. Mann. 2005. "Synthesis of Calcium Phosphate Nanofilaments in Reverse Micelles." *Chemistry of Materials* 17 (10):2765–70.

Sakamoto, Yasuhiro, Tae-Wan Kim, Ryong Ryoo, and Osamu Terasaki. 2004. "Three-Dimensional Structure of Large-Pore Mesoporous Cubic Silica with Complementary Pores and Its Carbon Replica by Electron Crystallography." *Angewandte Chemie International Edition* 43 (39):5231–34.

Salinas, A. J., A. I. Martin, and M. Vallet-Regí. 2002. "Bioactivity of Three CaO-P2O5-SiO2 Sol-Gel Glasses." *Journal of Biomedical Materials Research* 61 (4):524–32.

Salinas, A. J., J. M. Merino, F. Babonneau, F. J. Gil, and M. Vallet-Regí. 2007. "Microstructure and Macroscopic Properties of Bioactive CaO-SiO2-PDMS Hybrids." *Journal of Biomedical Materials Research Part B-Applied Biomaterials* 81B (1):274–82.

Salinas, A. J., M. Vallet-Regí, and I. Izquierdo-Barba. 2001. "Biomimetic Apatite Deposition on Calcium Silicate Gel Glasses." *Journal of Sol-Gel Science and Technology* 21 (1–2):13–25.

Salinas, Antonio J., and María Vallet-Regí. 2007. "Evolution of Ceramics with Medical Applications." *Zeitschrift für anorganische und allgemeine Chemie* 633 (11–12):1762–73.

Sanchez-Salcedo, S., D. Arcos, and M. Vallet-Regí. 2008. "Upgrading Calcium Phosphate Scaffolds for Tissue Engineering Applications." *Key Engineering Materials* 377:19–42.

Sanchez-Salcedo, S., M. Vila, I. Izquierdo-Barba, M. Cicuendez, and María Vallet-Regí. 2010. "Biopolymer-Coated Hydroxyapatite Foams: A New Antidote for Heavy Metal Intoxication." *Journal of Materials Chemistry* 20 (33):6956–61.

Sanchez-Salcedo, S., J. Werner, and M. Vallet-Regí. 2008. "Hierarchical Pore Structure of Calcium Phosphate Scaffolds by a Combination of Gel-Casting and Multiple Tape-Casting Methods." *Acta Biomaterialia* 4 (4):913–22.

Schmidt, S. M., J. McDonald, E. T. Pineda, A. M. Verwilst, Y. M. Chen, R. Josephs, and A. E. Ostafin. 2006. "Surfactant Based Assembly of Mesoporous Patterned Calcium Phosphate Micron-Sized Rods." *Microporous and Mesoporous Materials* 94 (1–3):330–38.

Serrano, M. C., M. T. Portoles, R. Pagani, J. S. De Guinoa, E. Ruiz-Hernández, D. Arcos, and M. Vallet-Regí. 2008. "In Vitro Positive Biocompatibility Evaluation of Glass-Glass Ceramic Thermoseeds for Hyperthermic Treatment of Bone Tumors." *Tissue Engineering Part A* 14 (5):617–27.

Sharp, K. G. 1998. "Inorganic/Organic Hybrid Materials." *Advanced Materials* 10 (15):1243–48.

Sharp, K. G., and M. J. Michalczyk. 1997. "Star Gels: New Hybrid Network Materials from Polyfunctional Single Component Precursors." *Journal of Sol-Gel Science and Technology* 8 (1–3):541–46.

Shirosaki, Y., K. Tsuru, S. Hayakawa, and A. Osaka. 2008. "Biodegradable Chitosan-Silicate Porous Hybrids for Drug Delivery." In *Bioceramics* (Vol. 20, Pts. 1 and 2), edited by G. Daculsi and P. Layrolle, 1219–1222. Stafa-Zurich: Trans Tech Publications.

Sing, K. S. W., D. H. Everett, R. A. W. Haul, L. Moscou, R. A. Pierotti, J. Rouquerol, and T. Siemieniewska. 1985. "Reporting Physisorption Data for Gas Solid Systems with Special Reference to the Determination of Surface Area and Porosity (Recommendations 1984)." *Pure and Applied Chemistry* 57 (4):603–19.

Soler-Illia, Galo J. de A. A., Clément Sanchez, Bénédicte Lebeau, and Joël Patarin. 2002. "Chemical Strategies to Design Textured Materials: From Microporous and Mesoporous Oxides to Nanonetworks and Hierarchical Structures." *Chemical Reviews* 102 (11):4093–4138.

Sun, L. M., C. C. Berndt, K. A. Gross, and A. Kucuk. 2001. "Material Fundamentals and Clinical Performance of Plasma-Sprayed Hydroxyapatite Coatings: A Review." *Journal of Biomedical Materials Research* 58 (5):570–92.

Tancred, D. C., B. A. O. McCormack, and A. J. Carr. 1998. "A Synthetic Bone Implant Macroscopically Identical to Cancellous Bone." *Biomaterials* 19 (24):2303–11.

Tsuru, K., S. Hayakawa, and A. Osaka. 2008. "Cell Proliferation and Tissue Compatibility of Organic-Inorganic Hybrid Materials." *Key Engineering Materials* 377:167–80.

Vallet-Regí, M. 2001. "Ceramics for Medical Applications." *Journal of the Chemical Society— Dalton Transactions* (2):97–108.

———. 2006. "Revisiting Ceramics for Medical Applications." *Dalton Transactions* (44):5211–20.

Vallet-Regí, M., and D. Arcos. 2005. "Silicon Substituted Hydroxyapatites: A Method to Upgrade Calcium Phosphate Based Implants." *Journal of Materials Chemistry* 15 (15):1509–16.

———. 2006. "Nanostructured Hybrid Materials for Bone Tissue Regeneration." *Current Nanoscience* 2 (3):179–89.

Vallet-Regí, M., D. Arcos, and J. Perez-Pariente. 2000. "Evolution of Porosity during in Vitro Hydroxycarbonate Apatite Growth in Sol-Gel Glasses." *Journal of Biomedical Materials Research* 51 (1):23–28.

Vallet-Regí, M., M. Gordo, C. V. Ragel, M. V. Cabanas, and J. San Roman. 1997. "Synthesis of Ceramic-Polymer-Drug Biocomposites at Room Temperature." *Solid State Ionics* 101:887–92.

Vallet-Regí, M., S. Granado, D. Arcos, M. Gordo, M. V. Cabanas, C. V. Ragel, A. J. Salinas, A. L. Doadrio, and J. San Roman. 1998. "Preparation, Characterization, and In Vitro Release of Ibuprofen from Al2O3/PLA/PMMA Composites." *Journal of Biomedical Materials Research* 39 (3):423–28.

Vallet-Regí, M., M. T. Gutiérrez-Ríos, M. P. Alonso, M. I. de Frutos, and S. Nicolopoulos. 1994. "Hydroxyapatite Particles Synthesized by Pyrolysis of an Aerosol." *Journal of Solid State Chemistry* 112 (1):58–64.

Vallet-Regí, M., I. Izquierdo-Barba, and F. J. Gil. 2003. "Localized Corrosion of 316L Stainless Steel with SiO2-CaO Films Obtained by Means of Sol-Gel Treatment." *Journal of Biomedical Materials Research Part A* 67A (2):674–78.

Vallet-Regí, M., I. Izquierdo-Barba, and A. J. Salinas. 1999. "Influence of P2O5 on Crystallinity of Apatite Formed In Vitro on Surface of Bioactive Glasses. *Journal of Biomedical Materials Research* 46 (4):560–65.

Vallet-Regí, M., J. Perez-Pariente, I. Izquierdo-Barba, and A. J. Salinas. 2000. "Compositional Variations in Time Calcium Phosphate Layer Growth on Gel Glasses Soaked in a Simulated Body Fluid." *Chemistry of Materials* 12 (12):3770–75.

Vallet-Regí, M., C. V. Ragel, and A. J. Salinas. 2003. "Glasses with Medical Applications." *European Journal of Inorganic Chemistry* (6):1029–42.

Vallet-Regí, M., A. Ramila, S. Padilla, and B. Munoz. 2003. "Bioactive Glasses as Accelerators of Apatite Bioactivity." *Journal of Biomedical Materials Research Part A* 66A (3):580–85.

Vallet-Regí, M., L. M. Rodriguez-Lorenzo, and A. J. Salinas. 1997. "Synthesis and Characterisation of Calcium Deficient Apatite." *Solid State Ionics* 101:1279–85.

Vallet-Regí, M., J. Roman, S. Padilla, J. C. Doadrio, and F. J. Gil. 2005. "Bioactivity and Mechanical Properties of SiO 2-CaO-P 2O 5 Glass-Ceramics." *Journal of Materials Chemistry* 15 (13):1353–59.

Vallet-Regí, M., A. M. Romero, C. V. Ragel, and R. Z. LeGeros. 1999. "XRD, SEM-EDS, and FTIR Studies of In Vitro Growth of an Apatite-Like Layer on Sol-Gel Glasses." *Journal of Biomedical Materials Research* 44 (4):416–21.

Vallet-Regí, M., A. J. Salinas, A. Martinez, I. Izquierdo-Barba, and J. Perez-Pariente. 2004. "Textural Properties of CaO-SiO2 Glasses for Use in Implants." *Solid State Ionics* 172 (1–4):441–44.

Vallet-Regí, M., A. J. Salinas, J. Ramirez-Castellanos, and J. M. Gonzalez-Calbet. 2005. "Nanostructure of Bioactive Sol-Gel Glasses and Organic Anorganic Hybrids." *Chemistry of Materials* 17 (7):1874–79.

Vallet-Regí, M., A. J. Salinas, J. Roman, and M. Gil. 1999. "Effect of Magnesium Content on the In Vitro Bioactivity of CaO-MgO-SiO2-P2O5 Sol-Gel Glasses." *Journal of Materials Chemistry* 9 (2):515–18.

Vallet-Regí, María, Isabel Izquierdo-Barba, and Montserrat Colilla. 2012. "Structure and Functionalization of Mesoporous Bioceramics for Bone Tissue Regeneration and Local Drug Delivery." *Philosophical Transactions of the Royal Society A: Mathematical, Physical and Engineering Sciences* 370 (1963):1400–1421.

Williams, D. F. 2008. "On the Mechanisms of Biocompatibility." *Biomaterials* 29 (20):2941–53.

Wolke, J. G. C., K. de Groot, and J. A. Jansen. 1998. "Subperiosteal Implantation of Various RF Magnetron Sputtered Ca-P Coatings in Goats." *Journal of Biomedical Materials Research* 43 (3):270–76.

Yabuta, T., E. P. Bescher, J. D. Mackenzie, K. Tsuru, S. Hayakawa, and A. Osaka. 2003. "Synthesis of PDMS-based Porous Materials for Biomedical Applications." *Journal of Sol-Gel Science and Technology* 26 (1–3):1219–22.

Yan, X. X., C. Z. Yu, X. F. Zhou, J. W. Tang, and D. Y. Zhao. 2004. "Highly Ordered Mesoporous Bioactive Glasses with Superior In Vitro Bone-Forming Bioactivities." *Angewandte Chemie—International Edition* 43 (44):5980–84.

Yanagisawa, T., T. Shimizu, K. Kuroda, and C. Kato. 1990. "The Preparation of Alkyltrimethylammonium-Kanemite Complexes and Their Conversion to Microporous Materials." *Bulletin of the Chemical Society of Japan* 63 (4):988–92.

Yao, J., W. Tjandra, Y. Z. Chen, K. C. Tam, J. Ma, and B. Soh. 2003. "Hydroxyapatite Nanostructure Material Derived Using Cationic Surfactant as a Template." *Journal of Materials Chemistry* 13 (12):3053–57.

Ye, F., H. F. Guo, and H. J. Zhang. 2008. "Biomimetic Synthesis of Oriented Hydroxyapatite Mediated by Nonionic Surfactants." *Nanotechnology* 19 (24). doi:10.1088/0957-4484/19/24/245605

Yu, C. Z., Y. H. Yu, and D. Y. Zhao. 2000. "Highly Ordered Large Caged Cubic Mesoporous Silica Structures Templated by Triblock PEO-PBO-PEO Copolymer." *Chemical Communications* (7):575–76.

Yuan, H. P., K. Kurashina, J. D. de Bruijn, Y. B. Li, K. de Groot, and X. D. Zhang. 1999. "A Preliminary Study on Osteoinduction of Two Kinds of Calcium Phosphate Ceramics." *Biomaterials* 20 (19):1799–1806.

Yuan, Z. Y., J. Q. Liu, L. M. Peng, and B. L. Su. 2002. "Morphosynthesis of Vesicular Mesostructured Calcium Phosphate under Electron Irradiation." *Langmuir* 18 (6):2450–52.

Yun, H. S., S. E. Kim, and Y. T. Hyeon. 2007. "Design and Preparation of Bioactive Glasses with Hierarchical Pore Networks." *Chemical Communications* (21):2139–41.

Yun, H. S., S. E. Kim, and Y. T. Hyun. 2008. "Fabrication of Hierarchically Porous Bioactive Glass Ceramics." In *Bioceramics* (Vol. 20, Pts. 1 and 2), edited by G.———. 2009. "Preparation of Bioactive Glass Ceramic Beads with Hierarchical Pore Structure Using Polymer Self-Assembly Technique." *Materials Chemistry and Physics* 115 (2–3):670–76.

Yun, H. S., S. E. Kim, Y. T. Hyun, S. J. Heo, and J. W. Shin. 2007. "Three-Dimensional Mesoporous-Giantporous Inorganic/Organic Composite Scaffolds for Tissue Engineering." *Chemistry of Materials* 19 (26):6363–66.

———. 2008. "Hierarchically Mesoporous-Macroporous Bioactive Glasses Scaffolds for Bone Tissue Regeneration." *Journal of Biomedical Materials Research Part B: Applied Biomaterials* 87B (2):374–80.

Yun, H. S., S. E. Kim, and E. K. Park. 2011. "Bioactive Glass-Poly (Epsilon-Caprolactone) Composite Scaffolds with 3 Dimensionally Hierarchical Pore Networks." *Materials Science & Engineering C:Materials for Biological Applications* 31 (2):198–205.

Zapanta-Legeros, Racquel. 1965. "Effect of Carbonate on the Lattice Parameters of Apatite." *Nature* 206 (4982):403–4.

Zhang, J. Y., Z. Luz, and D. Goldfarb. 1997. "EPR Studies of the Formation Mechanism of the Mesoporous Materials MCM-41 and MCM-50." *Journal of Physical Chemistry B* 101 (36):7087–94.

Zhao, D. Y., J. L. Feng, Q. S. Huo, N. Melosh, G. H. Fredrickson, B. F. Chmelka, and G. D. Stucky. 1998. "Triblock Copolymer Syntheses of Mesoporous Silica with Periodic 50 to 300 Angstrom Pores." *Science* 279 (5350):548–52.

Zhao, Y. F., and J. Ma. 2005. "Triblock Co-polymer Templating Synthesis of Mesostructured Hydroxyapatite." *Microporous and Mesoporous Materials* 87 (2):110–17.

Zhu, Y. F., C. T. Wu, Y. Ramaswamy, E. Kockrick, P. Simon, S. Kaskel, and H. Zrelqat. 2008. "Preparation, Characterization and In Vitro Bioactivity of Mesoporous Bioactive Glasses (MBGs) Scaffolds for Bone Tissue Engineering." *Microporous and Mesoporous Materials* 112 (1–3):494–503.

2 Mesoporous Ceramics as Drug Delivery Systems

In the last few years, the biopharmaceutical industries have been focusing their efforts on developing delivery systems for biomolecules, which can release different biopharmaceutical agents locally and in a sustained manner, at the same time as protecting them from degradation before reaching their target. Nowadays it is possible to deliver therapeutic dosages to selected areas of the body while side effects of a systemic dosage are minimized; that is, the concentration of the drugs at the precise sites is maintained within optimum range and under the toxicity threshold. The advances in technology allow the modification of drug release profile, adsorption, distribution, and elimination to improve the efficacy and safety, and also to tailor the treatment in response to the patient requirements.

These drug delivery systems should be made of a biocompatible carrier that allows high loading of biomolecules without any premature release of the cargo before reaching the target site, that is, zero premature release. In fact, there are several prerequisites for materials to be employed as delivery systems, such as (Slowing et al. 2008):

1. The material employed should be biocompatible.
2. The material should present high-loading or encapsulation abilities of the employed drug molecules.
3. There should be zero premature release to avoid any leaking of the drug molecules at the wrong site.
4. There should be certain control on the release of the drug molecules, with the adequate release rate to achieve an optimum concentration at the precise site.

Additionally, it is also desirable that certain cell or tissue specificity be achieved, especially in those cases where the drug is cytotoxic, such as in chemotherapy cancer treatments, and can present adverse side effects and limited effectiveness. In those cases, it is necessary to design a target-specific drug delivery system to transport an effective dosage of a certain drug to the targeted cells or tissues, as will be detailed later in this chapter.

In the recent past, several polymeric materials have been employed as delivery systems because they fulfill some of these requirements. In this sense, liposomes made of different phospholipids, polystyrene, dendrimers, or polylactic-polyglycolic acid have been used as drug delivery systems because they are biocompatible and biodegradable and they can encapsulate different drugs. However, it is very difficult to achieve zero premature release with these polymeric materials because as soon as they are in contact with aqueous solutions within the living body, they start biodegrading and subsequently leaking the encapsulated drugs. Additionally, the release rate of these polymeric materials is based on their degradation rate, so the composition must be modified to change the degradation and therefore achieve a proper release rate.

In the last few years, research in this area has been looking for structurally stable materials rather than soft degradable polymers, to develop drug delivery systems able to deliver a large number of drug molecules without any leaking before reaching the targeting site. Among those robust materials, mesoporous silicas have become very popular since they were proposed as drug delivery systems for the first time back in 2001 by the Vallet-Regí lab (Vallet-Regí et al. 2001).

In general, mesoporous silica materials fulfill all the requirements for a material to be employed as a drug delivery system:

1. Silica materials with defined structures and surface properties are known to be biocompatible (Slowing et al. 2008). In fact, silica has been selected to allow the biological application of several inorganic nanoparticles (Gerion et al. 2007; Bottini et al. 2007) and has also been added to artificial implants to induce osteogenic properties (Areva et al. 2007). In the same way, silica has also been employed in the formulation of different drug delivery systems, such as magnetic nanoparticles (Dormer et al. 2005), biopolymers (Allouche et al. 2006), or micelles (Huo et al. 2006).

2. The mesoporous network of cavities of mesostructured silica materials brings the possibility of introducing into these pores different cargos, such as drugs, peptides, proteins, or even other nanoparticles. The textural parameters of mesoporous materials have powered the high impact of these materials as drug delivery systems:

    The high pore volume permits the confinement of a large amount of a variety of pharmaceutical agents or biomolecules.

    Their large surface area confers to these materials a high potential for molecular adsorption.

    Their well-ordered mesopore distribution promotes the homogeneity and reproducibility of the adsorption and release mechanisms.

    Their high concentration of Si-OH groups at their inner and outer surface allows the modification of the surface chemistry through an easy chemical functionalization process, which favors a better control on the adsorption and release stages.

    The influence of all these parameters on the adsorption and release of the payload will be discussed in detail in this chapter of the book.

3. The possible organic functionalization of mesoporous silicas through an easy and quick process, as commented on in Chapter 1 of this book, brings the possibility of closing the pore entrances with diverse functionalities. In this way, the possible leaking of the entrapped drugs will be minimized, and the release will take place when the pore gates are opened in the so-called smart delivery systems.

4. As stated previously, the feasible functionalization approach allows the grafting of diverse organic moieties into the inner surface of the pores. Thus it is possible to control the interaction between the matrix and the drug by the careful selection of the functionalizing agent, and therefore the adsorption and, what is more important, the release kinetics of the drug. In this way, it is possible to release the cargo with a desired rate to achieve an effective drug concentration.

In addition to all these requirements, mesoporous silica materials are very popular as drug delivery systems because they are very easy to produce, as commented on in the previous chapter of this book, and their pore architecture can be easily tuned by the proper selection of the surfactant template. Another important characteristic of mesoporous silicas is that they are thermally and chemically stable, which permits carrying out several chemical reactions on their surface. This stability offers a great number of different engineering possibilities for these materials to be employed as drug delivery systems.

As stated earlier, the first time that ordered mesoporous silicas were proposed as drug delivery systems was back in 2001 (Vallet-Regí et al. 2001). The singular mesoporous structure of these materials inspired many important host-guest systems— among them, their use as carriers for drug delivery technologies (Figure 2.1). In this pioneering work, Vallet-Regí et al. designed an implantable material composed of MCM-41 with ibuprofen, an anti-inflammatory drug, confined in the mesopores. The pharmaceutical agent would then be locally released in bone tissue to reduce the inflammatory response after the implantation in a bone defect. Although this work was a conceptual piece of research, it was the first approach using ordered mesoporous materials as local drug delivery at the site where the action is required, and, additionally, controlling the drug concentration in plasma. Since the correct

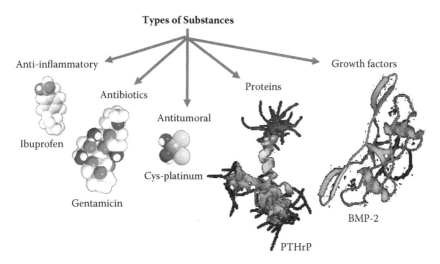

FIGURE 2.1 Schematic representation of the different biomolecules that have been loaded into mesoporous materials for drug delivery technologies.

localization and dosage of drugs are two milestones of drug delivery, these ordered mesoporous silica systems have been widely employed for these technologies, as will be detailed in this book.

2.1 ADSORPTION OF BIOMOLECULES

The drug-loading capacity of the different carriers is determined by many parameters, such as composition, particle size, or internal space for encapsulation. In the case of ordered mesoporous materials, the main factors controlling the adsorption capacity are the textural properties (Figure 2.2), such as pore size, mesostructure, and surface area, and the versatile chemistry thanks to the easy functionalization process.

Concerning the textural properties, the high pore volume of these materials, ca. 1 cm³/g, allows hosting a great number of pharmaceutical agents in their network of cavities. Their large specific surface area, ca. 1000 m²/g, offers a great potential for drug adsorption considering that the latter is a surface phenomenon. The highly ordered mesostructures, hexagonal or cubic arrangements, guarantee the homogeneity and reproducibility in the drug adsorption and release stages.

Regarding the tailorable surface with different chemical groups, the adsorption capacity can be modified on demand, favoring the retention of certain molecules through chemical interactions. The great versatility of the chemical functionalization allows an enormous variety of these interactions and therefore of the possible uploaded molecules.

Before starting to analyze the most important factors governing molecular adsorption into the mesoporous channels, the procedure employed to confine drug molecules into the mesopores, which is based on impregnation, should be revised. The first step should be the proper dehydration of the matrices, since the pores should be empty to allow molecules to be confined inside. To do so, the host matrices are

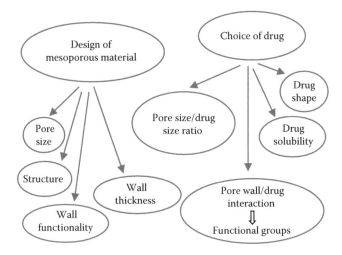

FIGURE 2.2 Main factors controlling the adsorption characteristics of ordered mesoporous materials.

normally dried in a vacuum oven at moderated temperatures, enough to evaporate the water but not very high so the mesostructure will not be affected by the process. After this, the mesoporous silica materials are soaked in a concentrated solution of the drug to be adsorbed. The solvent should be chosen depending on the solubility of the drug, as shown in Figure 2.3, but ideally aqueous solvents should be used to avoid any traces of toxicity from certain organic solvents. The concentration should be as high as possible to favor the drug impregnation on the silica walls. As a consequence, each type of drug presents optimum solvent and concentration to yield the largest possible number of physisorbed drug molecules. During the loading process, which normally takes over a day, magnetic stirring is applied to favor molecular diffusion into the channels, which is the final aim of the whole loading process.

The molecular confinement into the mesoporous cavities should be confirmed and quantified after the loading process, which can be done using several characterization techniques, as will be detailed throughout the next sections of this chapter. However, checking on the mesostructure survival after the loading process should be the first step, because if there are damages as a consequence of the loading conditions, there is no point in following the investigation. The mesostructure has to survive the loading process; this is a must in this technology. Thus small-angle X-ray diffraction (XRD) should be performed before and after the loading process, and the same diffraction maxima of the corresponding mesostructure should be obtained in

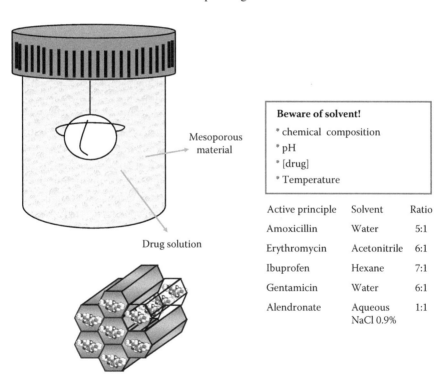

FIGURE 2.3 Schematic representation of the drug-loading process (left), and solvent selection depending on the drug to be adsorbed (right).

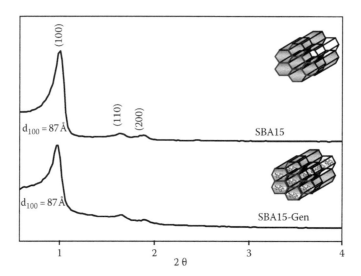

FIGURE 2.4 Small-angle XRD patterns of SBA-15 (top) and SBA-15 loaded with genta-micin (bottom). In both cases, the typical maxima of 2D-hexagonal symmetry are observed, that is, (100), (110), and (200).

both patterns, as observed in Figure 2.4, where small-angle XRD patters of SBA-15 before and after gentamicin loading are plotted (Doadrio et al. 2004). The same diffraction maxima should be observed to ensure that the mesostructure has survived the drug-loading process.

In the same way, nitrogen adsorption/desorption of the samples before and after drug loading should reveal similar isotherms, although the porosity and surface area values might be reduced as a consequence of the pore filling with the drug molecules, as observed in Figure 2.5. The same type of isotherms together with the small-angle XRD patterns should be good enough to confirm that the uptake process has not destroyed the organized mesostructure of these ordered mesoporous matrices. During those gentamicin adsorption studies, the difference in loading the drug into powder or disk pieces of mesoporous materials was evaluated (Doadrio et al. 2004). As expected, when SBA-15 materials were in powder form, there was more accessibility for the solvent to penetrate into the mesopores than in disks, which led to higher drug loading, as shown in the C, H, and N percentages from elemental analyses (Figure 2.5).

Additionally, it is also possible to confirm the prevalence of the ordered structure using transmission electron microscopy (TEM) of the loaded samples.

2.1.1 INFLUENCE OF PORE SIZE

Because size matters, the dimensions of both the molecule to be encapsulated and the channel diameter should be taken into consideration before taking the decision on which mesoporous host matrix should be employed. When the molecule to be adsorbed is larger than the pore entrance, the physical adsorption will take place

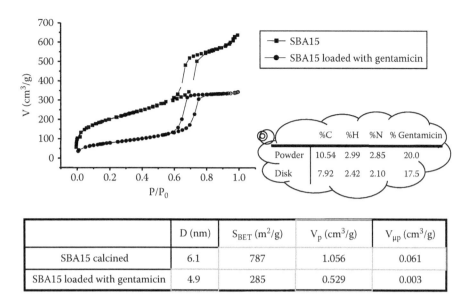

	D (nm)	S_{BET} (m^2/g)	V_p (cm^3/g)	$V_{\mu p}$ (cm^3/g)
SBA15 calcined	6.1	787	1.056	0.061
SBA15 loaded with gentamicin	4.9	285	0.529	0.003

FIGURE 2.5 Reduction on some textural parameters such as pore diameter (D), surface area (S_{BET}), and pore volume (Vp) of SBA-15 materials as a consequence of gentamicin loading. Additionally, as inset, elemental analysis of gentamicin-SBA-15 materials loaded as powder or as disk.

only at the external surface of the mesoporous material. However, when the targeted molecule is smaller than the pore diameter, the adsorption will take place in both the inner part of the mesopores and the external surface of the host matrix. In this sense, the pore diameter is known as a size-selective adsorption parameter.

Drug molecules constituted of simple organic groups are normally within the nanometer scale, with sizes ranging from a few to several dozen angstrom units. These small drug molecules would not have any problem entering the mesopore channels, so the physical adsorption would take place in the inner part of the pores and the external surface of the matrix particles. However, more complex metal-based drugs, which are used in certain cancer treatments to target DNA, such as platinum or ruthenium drugs, can reach bigger sizes due to the formation of organometallic complexes with different organic groups and therefore can have problems penetrating the mesopore entrances. Biomolecules, such as peptides or proteins, which are formed by the subsequent attachment of several amino acids, can reach huge dimensions, in terms of nanotechnology, which might limit their entrance into certain mesopores, and the loading might take place only at the external surface of the carrier matrix.

On the other hand, the mesopore diameters of the pore entrance might vary depending on the type of matrix employed. The most common ordered mesoporous matrices for drug delivery technologies are MCM-41 and SBA-15, which have in common their two-dimensional hexagonal arrangement. In those cases, the pore diameter is measured through the well-known surface characterization technique of N$_2$ adsorption/desorption analysis. This technique is based on the adsorption of N$_2$ gas on the surface of materials, and the measurement of the amount of gas adsorbed

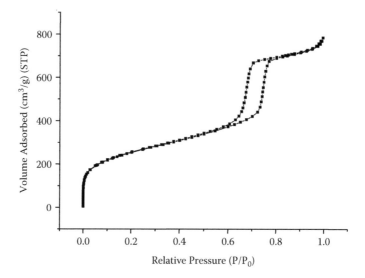

FIGURE 2.6 Typical adsorption/desorption isotherm of SBA-15 type ordered mesoporous materials.

over a range of partial pressures at a single temperature results in a graph known as an adsorption isotherm, as shown in Figure 2.6. Among the five types of adsorption isotherms established by the BDDT (Brunauer-Deming-Deming-Teller) method (Brunauer et al. 1940), porous adsorbents with pores in the radius range of approximately 15–1000 Å (mesopore range) present type IV isotherm.

In these types of mesoporous materials, the average pore diameter of primary mesopores can be obtained from the maximum of a pore size distribution using the BJH (Barret-Joyner-Halenda) method applied to the desorption part of the isotherm for certain relative pressures.

Once the dimensions of both guest molecules and host matrices are known, the selection of the correct matrix to be employed as drug delivery material should be taken. As mentioned previously, the pore diameter is one of the most important parameters, since it works as a size-selective adsorption parameter. In other words, the channel entrance size of mesoporous silica should be sufficiently large for the entrapment of the guest molecule. This fact was observed when trying to immobilize proteins in different mesoporous solids with different pore diameters (Yiu, Wright, and Botting 2001; Lei et al. 2004). The immobilization of proteins in mesoporous solids was triggered by the discovery of SBA-15, which presented extra large pore sizes up to 8–10 nm, and by the development of functionalization techniques, since the incorporation of organic groups on the surface is an excellent tool for the immobilization of biomolecules. The enzyme immobilization applying proteases, lipases, and peroxidases onto porous materials is used by the food and pharmaceutical industries (Yiu and Wright 2005) and has also been explored as substrate for protein separation (Zhao et al. 2002). In general, there are three main methods for enzyme immobilization inside the mesopores of the carrier matrices, as shown in Figure 2.7: (1) physical adsorption, (2) encapsulation, and (3) chemical binding.

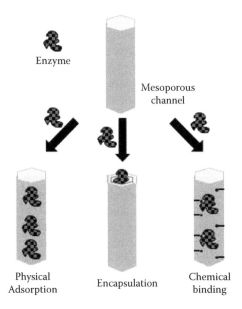

Enzyme

Mesoporous channel

Physical Adsorption

Encapsulation

Chemical binding

FIGURE 2.7 Three methods for enzyme immobilization in ordered mesoporous matrices: physical adsorption, encapsulation, and chemical binding.

In this section only the physical adsorption of the molecules will be discussed, since it is the simplest method for loading molecules into mesoporous materials: the host matrix is suspended in an enzyme solution overnight and then the loaded composite is recovered by filtration or centrifugation.

The size selectivity of the pore diameter was observed when testing different proteins, such as ovalbumin, serum albumin, conalbumin, cytochrome c, lysozyme, myoglobin, or β-lactoglobulin, into thiol-functionalized SBA-15 matrices. When the protein dimensions were smaller than the opening of the mesopores, the biomolecules were successfully physisorbed into the mesopore channels. However, when the tested proteins were larger than the mesopore diameter, such as in the case of ovalbumin, serum albumin, or conalbumin, the protein could not be physically adsorbed in the inner part of the mesopores and was adsorbed only at the external surface of the SBA-15 particles.

A similar experiment was carried out with ordered mesoporous glasses to show that the size-selectivity effect was independent of the composition of the host matrix (Menaa et al. 2010). The tested templated glasses presented similar textural parameters as the ordered mesoporous silicas, as has been detailed in Chapter 1 of this book, i.e., great porosity with uniform porous diameter and great surface areas. In this case, the diameter of the investigated mesoporous glasses, 4–5 nm, allowed the physical adsorption of apomyoglobin (4.1 nm) but impeded the loading in the inner part of the mesoporous cavities of albumin with larger size.

There are many methods to increase the pore diameter of ordered mesoporous silicas so they are able to adsorb large biomolecules. Among them, the most common procedure is the addition into the structure-directing template of certain swelling agents, such as 1,3,5-trimethylbenzene (Kruk, Jaroniec, and Sayari 2000), triisopropylbenzene (Kimura, Sugahara, and Kuroda 1998), or dodecane (Blin et al. 2000), which segregate to the hydrophobic pore of the templating micelles to enlarge them. The use of these agents can swell the pore volumes up to 30%, but the addition of high concentrations of them might result in a loss of the long-range order, which would lead to the formation of mesocellular foams, as will be detailed in the next section of this chapter. It is also possible to employ a mixture of surfactant blends with longer chains as structure-directing agents (Smarsly, Polarz, and Antonietti 2001). Even though great control on mesopore diameter can be achieved following this approach, the increment of the size of the channels is not good enough for physically adsorbing large proteins.

Our research group developed a straightforward approach to widen the pore diameter of SBA-15 materials through a hydrothermal treatment during the synthesis of the material (Vallet-Regí et al. 2008). Thus it was possible to control the pore size of SBA-15 by using different hydrothermal treatments, leading to mesopore diameters from 8.2 up to 11.4 nm after one and seven days of hydrothermal treatment, respectively, as shown in Figure 2.8. The target of this scientific work was to be able to load proteins with similar sizes to the mesopore diameter of traditional mesoporous materials. As a consequence of the enlargement of the mesopores after the thermal treatment for several days, it was possible to increase the BSA loading from 15%, assessed by the physical adsorption on the external surface of the material, to 27%,

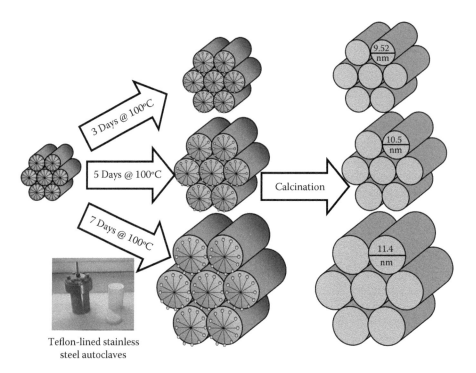

FIGURE 2.8 Hydrothermal synthesis of SBA-15 materials to increase the pore diameter. The longer the treatment, the larger the mesopore diameter of the produced materials.

which was the result of the physical adsorption on both the external surface and the inner area of the mesopores.

2.1.2 INFLUENCE OF SURFACE AREA

The surface area of the host matrix is a very important parameter since the physical adsorption of a molecule is a surface phenomenon, in which only the molecules that are in contact with the silica wall will be retained. The surface area of ordered mesoporous materials is determined by the application of the BET (Brunauer, Emmet, Teller) method to the nitrogen adsorption/desorption analysis, and gives an idea of how much exposed area a porous material has. The bigger this exposed area is, the greater the potential for molecular adsorption. In fact, ordered mesoporous materials are characterized by huge surface areas, in the order of 1000 m^2/g, which explains why the catalysis industries were the first researchers to produce these materials.

When these materials were proposed for drug delivery technologies, the inspiration came from those great values of surface area, which meant a high potential for drug adsorption. During the drug loading on the mesoporous matrix, and providing that the limiting factor of size-selective adsorption parameter is overtaken—i.e., the molecules are far smaller than the pore diameter—the drug molecules will enter the network of cavities of the matrix. However, a great number of molecules will not be retained inside the pores. Only some of them will actually interact with the pore walls and the rest will come out since there is not any attractive force to keep them inside. It is for this reason that the surface area is a key factor in the retention of the adsorbed particles, that is, in the loading process.

The best way to illustrate the importance of the surface area in the drug-loading process is comparing the loading of the same molecule into two matrices with different surface area. This comparative research was carried out loading alendronate, a potent bisphosphonate traditionally employed in the treatment of osteoporosis, into two matrices with the same mesostructure but different surface area, such as MCM-41 and SBA-15 (Balas et al. 2006), as observed in Figure 2.9. The original idea of this piece of research was to develop a drug delivery platform for alendronate, which has a poor intestinal adsorption and consequently requires high doses of oral administration with the subsequent side effects for the patients. The initial hypothesis consisted of the possibility of delivering bisphosphonates straight to the bone thanks to the introduction of the drug into osseous substitutes with drug delivery capabilities. In this sense, alendronate would have been delivered right at the place where it is specifically needed, rather than going through the whole digestive and circulating system. As mentioned above, the selected host matrices were MCM-41 and SBA-15, which present the same composition, pure silica, and mesostructure, 2-D hexagonal and p6mm symmetry, but different surface areas, 1157 m^2/g for MCM-41 and 719 m^2/g for SBA-15. Even though the selected host mesoporous materials present very different pore diameter, ca. 3 nm for MCM-41 and ca. 9 nm for SBA-15, the drug molecule was much smaller than both, ca. 0.6 nm. For this reason, the size-selective adsorption parameter did not have any effect on the different loading capacity; the loaded molecule was far smaller than the pore diameter. Alendronate adsorption was performed by soaking portions of the mesoporous materials in a concentrated

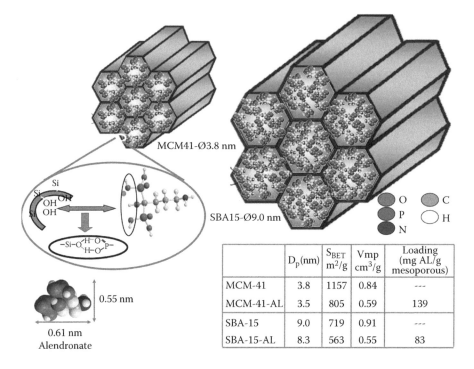

FIGURE 2.9 Alendronate loading into mesoporous materials with same mesostructure but different textural properties, such as MCM-41 (left) and SBA-15 (right). Similarly, alendronate-silica wall interaction is highlighted together with the dimensions of the drug molecule. In the right bottom corner, a table with the reduction on the textural parameters of the matrices as a consequence of drug loading can be observed.

solution of the drug buffered at pH 4.8 for 24 hours. After going through that loading process under the exact experimental conditions, the amount of bisphosphonate loaded was 14% for MCM-41 and 8% for SBA-15. These results showed the influence of the surface area on the molecular retention capacity, with greater drug loading in those matrices with higher surface area. As normally happens in science, the outcome of the research was far from the initial objectives, but the obtained conclusions were very important at the moment of the publication: the higher the surface area of the host matrix, the greater the loading capability.

The correlation of the loading amount of the drug with the BET surface area of mesoporous silica was also observed in MCM-41 materials prepared using templating surfactants of different alkyl chain length (Qu et al. 2006). The materials with higher surface area were able to load higher amounts of captopril, a popular drug employed for the treatment of hypertension and some types of congestive heart failure.

2.1.3 INFLUENCE OF PORE VOLUME

Pore volume has an important influence on the drug-loading capabilities when a high number of adsorbed molecules is required or when large molecules are uptaken. The pore volume in mesoporous materials is normally estimated from the nitrogen adsorption analysis according to the capillary condensation theory. In ordered mesoporous materials, the large mesopore volume, ca. 1 cm³/g, is one of the fascinating properties that inspired the molecular adsorption. In fact, the high pore volume allows the hosting of a large number of biomolecules into their mesostructure of cavities.

As previously mentioned, when high numbers of adsorbed biomolecules are required, pore volume, together with the previously detailed surface area, is a ruling factor in the drug amount that can be loaded. In this sense, the bigger the pore volume is, the larger the number of pharmacological agents that can be loaded. Depending on the final application of the drug carrier, sometimes high loading is necessary, and this can be achieved by repeating the loading process several times. Normally, it is good enough to suspend the mesoporous silica overnight in an aqueous solution of the drug to be adsorbed. If this process is repeated for several cycles, the consecutive impregnations would lead to the full pore volume filling. This phenomenon was observed when determining the loading capacity of ibuprofen as a model drug into MCM-41 carrier matrices (Charnay et al. 2004). Successive impregnations of MCM-41 particles were carried out with a solution of ibuprofen in ethanol, and after every step of the cycle, the pore filling was analyzed with nitrogen adsorption and X-ray diffraction. It was observed that the amount of drug loaded was increased after each step, and the intermolecular ibuprofen-ibuprofen interactions probably favored the significant improvement of the amount of ibuprofen loaded into MCM-41.

Regarding the possibility of loading large molecules, such as proteins, the available space for adsorbing the drug is equally as important as the pore diameter that would allow the entrance of the biomolecule into the channel. Among the diverse synthetic techniques to increase the pore volume of these mesoporous materials, perhaps the addition of a swelling agent such as 1,3,5-trimethylbenzene is the most popular. When adding this swelling agent to the synthesis of SBA-15 materials carried out with amphiphilic triblock copolymers as template, new materials called mesocellular silica foams (MCFs) are formed, which present increased pore size and pore volume (Lettow et al. 2000).

In fact, the pore size can be increased from 1.1 cm³/g in SBA-15 up to 1.9 cm³/g in MCF. This huge increment in pore diameter allowed the introduction of large proteins, such as bovine serum albumin (BSA) (Schmidt-Winkel et al. 1999), to be then released, as represented in Figure 2.10.

The influence of the pore volume was observed when quantifying the amount of BSA loaded into the network of cavities: the higher the pore volume, the greater the protein adsorption. Thus BSA loading was increased from 15% in traditional SBA-15 materials up to 24% in these new MCF materials (Vallet-Regí et al. 2008).

As mentioned previously, when the physically adsorbed molecule is larger than the pore entrance and/or the pore volume is not large enough to accommodate it, that molecule would be adsorbed only at the external surface of the host particles. However, when the molecule is small enough to go through the pore entrance, it is supposed to be physically confined in the inner part of the channels. In these cases, it is necessary to confirm that the drug molecules are confined inside the mesopores and not only on the outer surface of the particles. Traditionally, nitrogen adsorption analysis has been employed to solve this question because it is possible to obtain the

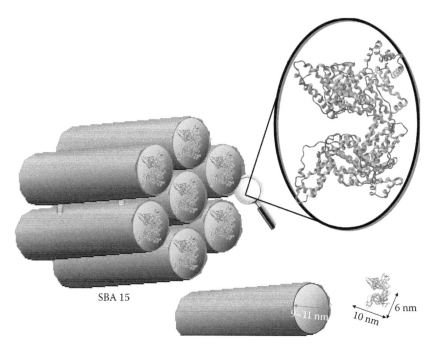

SBA 15

~11 nm

6 nm

10 nm

FIGURE 2.10   Representation of SBA-15 materials with enlarged pore diameter and pore volume to be able to load proteins such as bovine serum albumin.

surface area, pore size distribution, and pore volume of host matrices before and after the loading process. Even though the reduction on the surface area after uptaking the drug is indicative of molecular adsorption, it does not reveal if this reduction is exclusively at the inner surface of the mesopores. However, the reduction on the available pore volume after the loading process is, with no doubts, indicative that the molecules are filling the pore space, which means that drug molecules are being confined inside the mesopores.

However, nitrogen adsorption/desorption analysis is an indirect way of characterizing the filling of the mesopores with the drug molecules. The reduction of the available pore volume after the loading process is indicative that the drug molecules are occupying this space, but it is not direct evidence of which molecules are in the pores. Fourier transform infrared (FTIR) spectroscopy can be employed to identify those molecules, applying this technique before and after the loading process. The vibration bands corresponding to the drug, which would vary depending on the functional groups of the adsorbed molecule, should be observed in the spectrum after the loading, confirming the presence of the drug into the silica matrix. However, FTIR does not allow one to elucidate if the adsorbed drug molecules are inside or outside the pores. The same drawback was observed with thermogravimetry, elemental analysis, and X-ray fluorescence (XRF), which are traditionally employed for quantifying the number of adsorbed molecules but make no differentiation on the external or internal adsorption.

This scenario changed in 2010, when our research group demonstrated for the first time that it was possible to detect drug molecules confined in the inner part of pore channels or ordered mesoporous matrices (Vallet-Regí et al. 2010). This was possible thanks to the advances in electronic microscopy—specifically, using scanning transmission electron microscopy (STEM) with a spherical aberration (Cs) corrector. This technique is based on scanning a sample using an electron probe that is focused down to 1 nm on the specimen. Then the collected scattered electrons in each probe position by the high-angle annular dark-field (HAADF) detector at the bottom of the sample in synchronism with the scanning probe are employed to form the STEM images. In this way, when equipping an atomic resolution analytical microscope with a STEM Cs corrector, it is possible to perform amazing microscopy with enough resolution to achieve atomic-level analysis. This technique was employed to analyze a SBA-15 matrix loaded with zoledronate, another potent bisphosphonate drug, similar to alendronate. A microscope equipped with a STEM Cs spherical aberration corrector and electron energy loss spectroscopy (EELS) can be employed to show the distribution of silicon, oxygen, nitrogen, and carbon throughout the silica network. In other words, the high resolution of the microscope allows the differentiation between pore domains and the silica wall, which makes this technique a perfect tool to find out where the molecules are being confined. Apart from confirming the ordered mesostructure of the matrix, the HAADF-STEM images taken when the electron beam illuminated the wall confirm that the silica walls are composed of Si and O, as was expected. When the electron beam was oriented in the parallel direction of the mesopores, the energy loss spectra showed that the material located inside the mesopores presented carbon and nitrogen in its composition, which are unequivocally from the confined organic molecule zoledronate. The obtained data confirmed that the drug molecules were confined in the inner part of the mesopores of SBA-15 host matrices.

2.1.4 Influence of Functionalization

The first consideration that should be taken when dealing with ordered mesoporous silicas as drug delivery carriers is that molecular adsorption, i.e., drug loading, is a surface phenomenon, as mentioned in Section 2.1.2. In this sense, the retention of the adsorbed molecules in the inner part of the mesopores and the external surface of the particles is governed by the chemical interactions between Si-OH groups from the silica host and the functional groups from the guest molecules. It seems quite straightforward that to modify the type of the host-guest interaction, a modification on the silica walls should be performed. Depending on the desired modification and the chemical groups of the molecule to be adsorbed, the walls of the silica matrix can be functionalized with many different chemical groups. Certainly, it is possible to design the mesoporous matrix to improve the adsorption capacity and/or retard the release kinetics of the adsorbed molecules.

As commented on in Chapter 1 of this book, the high density of silanol groups within these silica mesoporous materials makes possible their easy modification through the grafting of alkoxysilanes with organic groups. Through this method, the chemical properties of the host matrices can be tuned at convenience, so there would be a stronger host-guest interaction leading to a higher retention of the adsorbed molecule. This would mean a greater drug adsorption and a more sustained release of the drug, both factors being important milestones in drug delivery technologies.

The functionalization reaction of silica materials with organic alkoxysilanes is normally quantitative, and there are many characterization techniques that can be employed to: (1) confirm the correct anchoring of the modifying agent, such as ^{29}Si NMR and FTIR (Horcajada et al. 2006; Menaa et al. 2009), and (2) quantify the degree of modification, such as elemental analysis, thermogravimetry, or X-ray photoelectron spectroscopy (XPS) (Menaa et al. 2009; Balas et al. 2008).

In any case, it is important to highlight that the selection of the organic modification should be made depending on the functional groups of the molecule to be adsorbed. The number of possible organic modifications is almost infinite, which confers a great versatility to these carrier materials (Manzano and Vallet-Regí 2010).

The strong influence of the functionalization of the matrix over the loading process was observed for the first time in the pioneering work of ibuprofen adsorbed into MCM-41 (Vallet-Regí et al. 2001). Before any organic modification, the interaction of the pharmaceutical agent with the silica walls relied on the relationship between the carboxylic acid group from the drug, COOH, and the silanol groups from the silica walls, Si-OH, as represented in Figure 2.11.

Regarding the possibility of modifying the chemistry of the silica walls, researchers thought about increasing the strength of the host-guest interaction through the functionalization with amine groups. The aim of this approach was promoting a stronger host-guest interaction to favor the molecular adsorption and delay the release of the drug molecules. Effectually, the more powerful interaction between the carboxylic acid group from ibuprofen and amine groups from the modified MCM-41 matrices led to an increase in the ibuprofen adsorption and promoted more sustained drug release kinetics than with unmodified matrices. This was the first time that some kind of control on the drug delivery was achieved, as will be detailed in the next sections of this chapter.

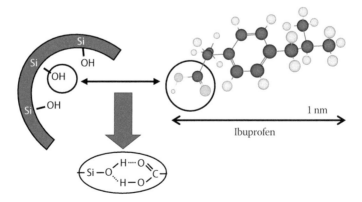

FIGURE 2.11 Schematic representation of the host-guest interaction that takes place between silanol groups from the silica walls and carboxylic acid from ibuprofen molecules.

Organic modification has been revealed as the most important factor affecting the drug loading on ordered mesoporous materials. It is true that the pore size is a limiting factor, and the pore volume needs to be large enough to confine the adsorbed drug molecules. It is also true that molecular adsorption is a surface phenomenon, and therefore when surface area is higher, the chances of obtaining larger molecular adsorption are greater. However, the modification of the pores' surface has been the most important variation to increase the drug-loading capability of these carrier matrices. The clearest example of this dependence was found when loading alendronate into different mesoporous matrices, MCM-41 and SBA-15 (Balas et al. 2006). When loading alendronate into pure silica materials, 14 and 8 wt% of drug confinement were obtained for MCM-41 and SBA-15, respectively. However, after modifying the surface of the mesoporous materials with amine groups, alendronate loading was increased up to 37 and 22 wt%, respectively, because of the attractive interactions between the amine groups from the host matrix with the bisphosphonate groups of the guest molecule, as represented in Figure 2.12. This difference gives an idea of how important the functionalization could be, improving almost three times the loading capacity of ordered mesoporous matrices.

The degree of functionalization, i.e., the number of organic moieties anchored to the surface of the walls, will play an important role in the number of adsorbed molecules. It is logical to think that the more organic the groups, the more points for molecular interaction with the drug, so the more adsorption capacity of the matrix. This is true, but only up to a level where it is not possible to absorb more drug molecules. This can be experimentally observed when performing a gradual functionalization of a silica matrix with amine groups to then load alendronate molecules. The gradual coverage of the matrix surface can be achieved using the post-synthesis

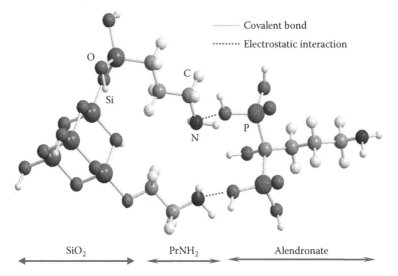

———— Covalent bond

·········· Electrostatic interaction

SiO_2 $PrNH_2$ Alendronate

FIGURE 2.12 Schematic representation of the host-guest interaction that takes place between amine groups from the functionalized matrix walls and phosphonate groups from alendronate molecules.

method and adjusting the amount of amino-silane employed. The first time that this technique was explored, different degrees of surface functionalization were achieved, from 25% up to 100% (percentages based on the total amount of silanol present in the matrix surface) (Nieto et al. 2008). Consequently, there was a gradual loading of alendronate, with more alendronate loaded into those matrices with more amine groups grafted. In general, and up to a certain functionalization level, the larger the content of grafted functional groups, the greater the amount of drug loaded into the mesopore channels of the matrices.

Over the years, the design of specific matrices through organic functionalization of mesoporous materials to improve adsorption capacity or release kinetics has been revealed as one of the most important parameters to take into account. Properly tuning the chemical properties of these mesoporous carriers has even been a prerequisite to allowing molecular adsorption. This is the case with adsorption of tryptophan, an essential amino acid that is present in many structural or enzyme proteins, into SBA-15 materials (Balas et al. 2008). If the silica matrix is not previously functionalized with certain groups, it is not able to uptake any tryptophan into the mesopores due to the hydrophobic nature of this amino acid. However, an organic modification of the silica walls of SBA-15 with quaternary alkyl amines (Figure 2.13) changes the hydrophobicity of the carrier surface, which promotes the tryptophan adsorption into the mesopores.

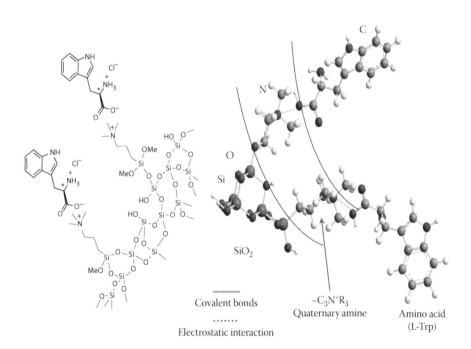

FIGURE 2.13 Schematic representation of the host-guest interaction that takes place between quaternary alkyl amine groups from the functionalized matrix walls and carboxylic groups from tryptophan amino acid.

The increase of the amino acid adsorption is explained by the strong interactions that take place between the hydrophobic alkyl chains and the indole group from the tryptophan. In fact, almost two-thirds of the silica surface area functionalized, which radically changes the hydrophobic character of the carrier and hence increases the amino acid adsorption into the mesopores.

However, all these functionalization approaches do not guarantee that the loading of the molecules is exclusively into the inner volume of the pores. There is a high probability that some of the adsorbed molecules will be located at the external surface of the carrier particles, where they are not protected from the external environment, which might lead to degradation or denaturation. A possible approach to avoid this drawback and to ensure that the adsorbed molecules are exclusively confined inside the mesoporous cavities is the selective functionalization of the external surface, which would impede external molecular adsorption and will leave the internal pore volume available for drug molecules' confinement. This strategy has been adapted from the catalysis industry, where MCM-41 has been widely employed as catalyst support (Shephard et al. 1998), in which derivatization of the surface with the appropriate organic moiety takes place predominantly on the external surface of the particles because of its greater accessibility. The same objective can be achieved through the functionalization of both the internal and external surfaces with fluorenylmethoxycarbonyl (Fmoc)-modified organosilanes, and the subsequent selective deprotection of the external surface groups, which can be further functionalized (K. Cheng and Landry 2007). Then the internal Fmoc groups can be deprotected and the molecular adsorption would occur only at the inner part of the pores. Other strategies for selective functionalization of mesoporous materials include the sequential co-condensation approach, in which functional groups can be completely dispersed inside the mesoporous channels, concentrated in parts of the mesopores, or exclusively placed on the external surface depending on the time of addition during the synthetic process (Kecht, Schlossbauer, and Bein 2008). However, the simplest way of selectively functionalizing the external surface of mesoporous materials is based on the reaction of organosilanes with mesoporous silica still containing the structure-directing agent. Once the functionalization takes place, the template can be removed by soft methods, such as solvent extraction, and the mesopores will be freely available for molecular adsorption (Gartmann and Bruhwiler 2009).

The increased loading effect due to the functionalization is not exclusively for pure silica materials. Bioactive mesoporous glasses can also be modified with hydrophobic groups to incorporate ipriflavone, which is a highly hydrophobic antiosteoporotic drug (Lopez-Noriega, Arcos, and Vallet-Regí 2010). Before any organic modification of the mesoporous glass, the amount of drug loaded is very low, ca. 13 mg/g. However, the loading capacity of the same bioactive mesoporous glasses was increased after modifying their surface with mercaptopropyl groups, ca. 40 mg/g; aminopropyl groups, ca. 61 mg/g; hydroxypropyl groups, ca. 60 mg/g; and phenyl groups, ca. 120 mg/g. In all functionalization approaches there are hydrogen interactions with the drug that increase the adsorbed quantity. Additionally, in the case of phenyl groups, there are pi-pi staking interactions that increase the strength of the interaction and lead to higher drug adsorption, which highlights the importance of the selection of the organic group to perform the functionalization.

2.2   RELEASE OF BIOMOLECULES

Drug delivery systems are expected to deliver their cargo with a careful controlled release of the previously adsorbed pharmaceutical agents or biomolecules. Most drug delivery systems can be divided into two groups: passive type, where the cargo release rate is governed by processes such as molecular diffusion from a stable porous mesostructure or dissolution in degradable carriers; and active type, where the cargo is released in response to an external or internal stimulus that acts as trigger, in the so-called smart drug delivery systems (Arruebo, Vilaboa, and Santamaria 2010). In this section of the chapter, the diffusion of the drugs through the mesostructured pores will be covered, while the active systems such as smart delivery devices will be detailed in the next chapters of this book.

The methodology for studying the release of the previously physisorbed molecules starts with the release experiments in vitro. The drug-loaded mesoporous matrices are normally compacted into pellets using unidirectional and isostatic presses, so the material can be easily handled. Those silica pellets are soaked in a simulated body fluid (Kokubo and Takadama 2006), which is a solution that mimics the human plasma, at 37°C and they are stirred at low revolutions per minute using an orbital shaker, as shown in Figure 2.14. In this way, the physiological conditions that these

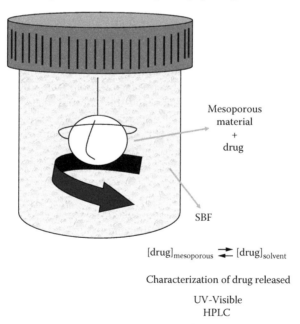

FIGURE 2.14   Schematic representation of the procedure for the in vitro drug release experiments.

materials would find in their final use as drug delivery systems are emulated. During the release experiments, the drug molecules would diffuse out of the mesopores and be released to the surrounding media. Subsequently, small quantities of the soaking solution with the released drug dissolved in it are taken at determined times, and the drug concentration on these aliquots is quantified using ultraviolet (UV) spectroscopy and/or high-performance liquid chromatography (HPLC).

The obtained cumulative release is commonly represented as a function of time, and the release pattern obtained from this plotting would determine the efficiency of the designed carrier system.

In most of the release patterns of these ordered mesoporous materials, there are two distinct areas: (1) an initial burst of drug release because of the release of the drug adsorbed at the surface of the surface of the particles that is quickly released to the delivery medium, and (2) a more sustained release due to the slow diffusion of the drug molecules along the mesoporous channels to the surrounding medium.

It is well known that correlations of in vitro and in vivo drug release systems are uncommon (Kohane and Langer 2010), since in an in vivo situation there are many different external parameters that can influence the release kinetics. However, the in vitro results would mainly depend on the designed system itself, so the obtained data from these curves can help to improve and implement the designed carrier system. Once this first testing has been done and the system has been optimized, the next logical step in the research could be performed with cell cultures and/or animal tests. In the next few sections of this chapter, the factors affecting the in vitro release will be detailed first; then there will be a comment on Chapter 5 regarding the studies performed with both cell cultures and different animal models when dealing with the in vitro/in vivo behavior of these matrices.

2.2.1 INFLUENCE OF PORE SIZE

Besides the influence of the textural properties of mesoporous materials on the molecular adsorption of many drugs, their influence on the diffusion of the adsorbed drug molecules through the mesopores and consequently on the release kinetics is also very important. The pore size, which was a size-selective parameter in the loading process, can be employed as molecular release rate modulator. To demonstrate this rate modulator effect, MCM-41 matrices with different pore diameters can be produced employing surfactant templates with different alkyl chain lengths. The direct consequence of employing cationic surfactants with longer or shorter chains as structure-directing agents is the production of ordered mesoporous materials with the same mesostructure but different pore diameters. When testing these materials as ibuprofen release systems, Figure 2.15 shows the direct relationship between the pore size and the release rate: A decrease on the pore size led to a decrease on the release rate of ibuprofen (Horcajada et al. 2004). As a consequence, it is clear that release kinetics of ordered mesoporous materials can be modulated by the proper tailoring of the mesopore size.

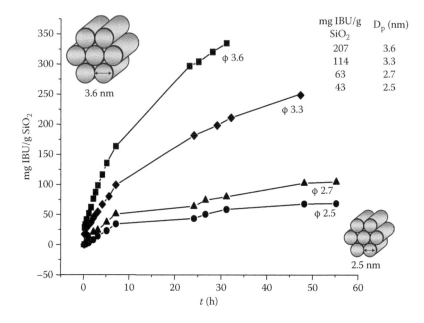

FIGURE 2.15   Ibuprofen release patterns from MCM-41 materials with different mesopore diameter.

2.2.2   INFLUENCE OF SURFACE AREA

In the previous section of this chapter, when the drug-loading procedure was tackled, it was highlighted that molecular adsorption in ordered mesoporous materials is a surface phenomenon. The molecules are confined inside the pores through the attractive interaction with the internal mesopore walls. It is reasonable to think that the inverse process, the release of the adsorbed drugs, would also depend on the surface area of the material. Effectually, different experiments have shown that when the surface area of a mesoporous material is very high, apart from presenting a great molecular retention, there is a slower drug release in comparison with materials with smaller surface area. The reason for this is that in the diffusion process throughout the mesopores, the molecules on their way out would find more available area that can interact with them, promoting extra host-guest interactions that retard the release kinetics.

The influence of the surface area on the release of the guest molecules can be so important that they can even modify the release patterns. This effect can be observed when studying the release of the same drug, alendronate, from two matrices with the same mesostructure but different surface area, such as MCM-41 and SBA-15 (Figure 2.16). The release of the pharmaceutical agent from MCM-41, with a specific surface area of 1175 cm^2/g, presents first-order kinetics, while the release of alendronate from SBA-15, with a specific surface area value of 719 cm^2/g, is observed to follow linear or zero-order kinetics (Balas et al. 2006).

	Loading mg AL/gSiO$_2$	Delivery AL in 24h (mg AL/g SiO$_2$)	Total Delivery time (h)
MCM-41	---	---	---
MCM-41-AL	139	80.7	72
NH$_2$-MCM-41	---	---	---
NH$_2$-MCM-41-AL	366	102.9	264*
SBA-15	---	---	---
SBA-15-AL	83	45.7	264
NH$_2$-SBA-15	---	---	---
NH$_2$-SBA-15-AL	220	24.2	264**

* Incomplete delivery (76%) ** incomplete delivery (69%)

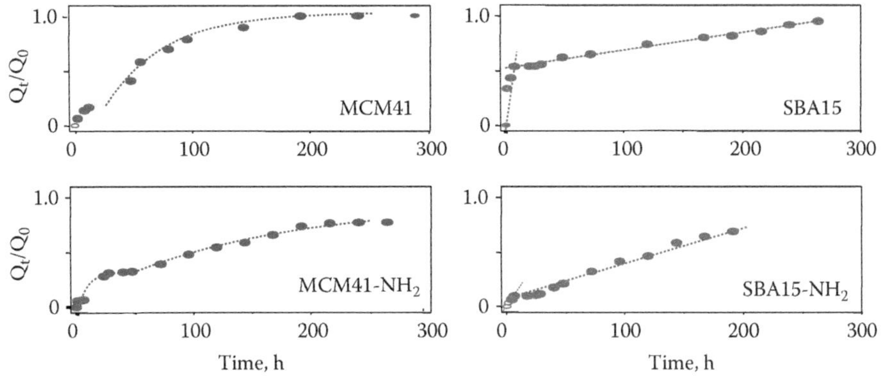

FIGURE 2.16   Alendronate delivery from ordered mesoporous matrices with different surface area: MCM-41 and SBA-15.

2.2.3 Influence of Functionalization

In the same way that organic modification of the silica walls was the most determining parameter in the molecular adsorption into these mesoporous materials because the retention of the drug molecules could be enhanced, the release is going to be equally affected. In fact, by using the appropriate functionalization it is possible to control the release of the adsorbed molecules, because the host-guest interaction can be controlled as required. This control on the release has been observed in all types of ordered mesoporous silica materials, independently of the mesostructure and the textural parameters, because the functionalization can be specifically selected to interact with the chemical groups of the drug, promoting a better retention of this drug into the matrix.

In the previously commented on case of testing two different matrices, such as MCM-41 and SBA-15 materials, with the same drug, alendronate, the surface area seemed to be a key factor in both drug adsorption and release. However, when those matrices were organically modified with amine groups, selected to enhance the interaction with phosphonate groups from alendronate, in addition to increasing the drug adsorption, the release rate was reduced. Both effects are produced by the same cause. There is a stronger host-guest interaction between amine and phosphonates than silanol and phosphonates. As a consequence, both molecular adsorption and drug release are improved through a better control. In fact, when the release patterns are compared with those from unfunctionalized materials, it is clear that the organic functionalization is retarding the release, which could be of great value when sustained release is required.

A similar effect is observed when proteins are loaded into functionalized matrices such as SBA-15 and MCF. Although there is a greater BSA uptake into MCF because of the larger pore diameter and higher porosity than SBA-15, the release patterns of BSA from both unmodified matrices is similar. However, when the matrices are functionalized with amine groups, the interaction between the protein and the amine-modified mesoporous carriers is stronger. This attractive interaction promotes a decrease in the burst effect and a much more sustained release of the protein from the carriers. Thus through an easy organic modification with amine groups, it is possible to achieve certain control on the release of a protein from ordered mesoporous materials.

The functionalization influence has been observed to be more important than the surface area effect, the mesopore diameter and volume influence, and even the mesostructure. To prove this, two mesoporous materials with different mesostructure, such as MCM-41 and MCM-48, can be loaded with the same antibiotic, erythromycin. Although the release patterns might look slightly different (Figure 2.17), the real change comes when both matrices are functionalized with long alkyl chains to change the hydrophobic character of the mesopore walls. In those alkyl-modified matrices, the antibiotic release patterns are very different from the unmodified. There is a retention of the erythromycin in the pores because of the favorable interaction with the more hydrophobic surface, which leads to a more sustained release of the drug.

An imaginative approach to control the number of drug molecules adsorbed and, what is more important, the release kinetics of these drug molecules, is the covalent grafting onto the surface of mesoporous matrices of dendrimers, which are highly branched monodisperse macromolecules that might present many different functionalities. In the previous section of this chapter, when evaluating the influence of the functionalization over molecular adsorption the possibility of gradually functionalizing the surface to control de-adsorption of the drug was mentioned. That control was partially possible at low functionalization degrees, but when increasing the number of organic moieties, the control was lost. This is due to the limited number of silanol groups available at the surface of the materials, so once they are all functionalized, the excess of functionalizing agents is not going to produce any effect and they will go away during the washing process. However, the use of different generations (first, second, and third) of poly(propyleneimine) dendrimers allows for having many more amine groups (4, 8, and 16 primary amine groups on their periphery, respectively) at the inner part of the mesopores readily available to attract drug molecules than conventional amine-silane grafting, as can be observed in Figure 2.18. These different generations of amine dendrimers can be easily grafted to the surface of the channels of SBA-15 through the introduction of alkosysilanes into the dendrimers' structure (Gonzalez et al. 2009). When evaluating the ibuprofen-loading capacity, the amount

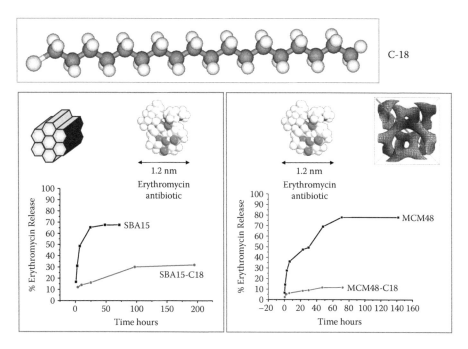

FIGURE 2.17 Erythromycin release patterns from ordered mesoporous matrices with different mesostructure: SBA-15 and MCM-48, both nonmodified and functionalized with C18 alkyl chains.

of confined drug increases with each dendrimer generation, because of the greater number of available amine groups on the second and third generation. The amount of loaded ibuprofen ranges from 22% in pure SBA-15 material to 29, 41, and 48% for SBA-15 modified with first-, second-, and third-generation dendrimer, respectively. However, what is really interesting from the perspective of a possible application in the clinic is the reduction of the release rate of ibuprofen when dendrimers are employed. Depending on the dosage required for the patient, the appropriate dendrimer generation should be employed to gather an effective control over the release kinetics of the employed drug.

Over the years of experience using these matrices as drug delivery systems, the organic functionalization has been observed as the most relevant factor in controlling drug release kinetics. It is for this reason that during the designing process of a drug delivery system using ordered mesoporous materials, the functionalization with determined organic groups must be carefully considered and selected depending on the biologically active molecule to be confined into the mesoporous channels.

One important tool for the designing process of a drug delivery system is the development of a prediction method on how the release will be depending on the physicochemical characteristics of the carrier matrix. This original approach seems to be a powerful tool for researchers in this area, since it allows the design of delivery vectors with predicted release kinetics, with the subsequent benefits of saving time and resources. This modeling has been performed for the release of zoledronate from

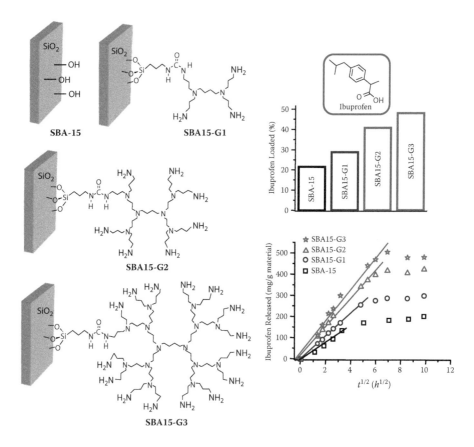

FIGURE 2.18 Schematic representation of different generations (G1, G2, and G3) of dendrimer grafting onto the surface of SBA-15 (left), and ibuprofen adsorption (top right) together with ibuprofen release patterns from those matrices (bottom right).

SBA-15 materials, and the experimental results have been found to validate the prediction made (Manzano et al. 2011), as can be observed in Figure 2.19.

The development of a modeling system is a very useful tool for the designing of implantable drug delivery systems allowing the prediction of the release kinetics of a biomolecule with anticipation. Regardless of the matrix and the drug, this prediction can be applied to any kind of mesoporous system and any biomolecule such as peptides or proteins. Consequently, the research capabilities would be more efficiently applied predicting release kinetics in advance, with the subsequent savings on time and resources.

2.2.4 STABILITY OF ORDERED MESOPOROUS MATERIALS

The interaction of synthetically produced ordered mesoporous materials with the physiological environment is an important point to consider for biomedical applications. In this sense, and depending on the final application of ordered mesoporous

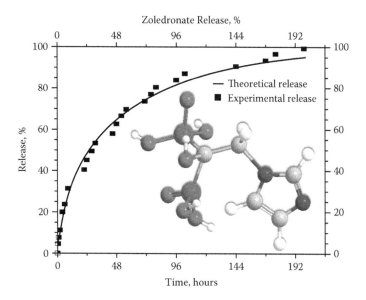

FIGURE 2.19 Theoretical and experimental zoledronate release patterns from SBA-15 materials.

materials, sometimes a degradable platform is preferred, such as in the case of bone repair or replacement materials, and in other cases stable platforms are needed, such as drug delivery vectors. The former will be discussed in Chapter 5 of this book, while the latter, stability of drug delivery applications, will be detailed here.

The potential degradation of the carrier matrix under physiological conditions is an important parameter that needs to be controlled in all drug delivery systems for two main reasons:

1. If the degradation of the matrix is fast, there will be two competitive release mechanisms: (a) *diffusion-controlled delivery*, which consists of the diffusion of the drugs throughout the mesopore channels, and (b) *degradation-controlled mechanism*, in which the release of the drug molecules takes place as a consequence of the erosion of the matrix. If these competitive mechanisms take place at the same time, there will be a complete lack of control on the release kinetics, which should be avoided.
2. Even if the carrier matrix is perfectly biocompatible, the degradation products might be toxic, which should be avoided for in vivo applications.

The response in biological systems and consequently the degradation behavior of ordered mesoporous materials will be controlled by their compositional and textural properties. To analyze the dynamic response of mesoporous materials under relevant biological conditions, it is possible to use thin films of mesoporous oxides as a model system to study the dynamic nature of the interaction (Bass et al. 2007). The use of mesostructured oxide thin films as model systems rather than bulk or nanoparticles materials is due to the relatively easy route for producing them and, more importantly, the convenient geometry that they present for the study of the evolution of textural parameters as they interact with physiological fluids. The dynamic behavior of these materials was found to be dependent on their composition, porosity, and calcination temperature. The complete degradation of silica films was in the same time scale as the attachment and spreading of many types of mammalian cells, and this degradation resulted in dissolution down to molecular species. There are many different approaches for increasing the stability of mesoporous materials in physiological media (Sanchez et al. 2008), and among them the most interesting from the drug delivery perspective include the organic functionalization of the silica walls and the production of mesoporous oxides with other metals, such as alumina or zirconia.

The reduction in matrix degradation under physiological conditions has been observed not only in mesoporous thin films but also in bulk mesoporous materials. The degradation of SBA-15 bulk particles was evaluated under different physiological solutions, such as saline solution, SBF, and Dubecco's modified Eagle medium (DMEM) (Izquierdo-Barba et al. 2010). Independently of the tested media, the lixiviation curves (degradation of SiO_2) are characterized by a relatively fast dissolution rate up to ca. 250 hours (Figure 2.20). The mesostructure of the matrix was analyzed by TEM, and the micrographs showed that after 10 days the ordered arrangement of the mesopores was still there, although after 60 days it was almost gone in some domains with 2D hexagonal structure.

However, when SBA-15 materials were functionalized with organic groups, such as amine, methyl, and octyl organosilanes, the amount of silica dissolved was reduced in comparison with pure silica materials because the organic coverage protects the matrix from the nucleophilic attack of water. As can be observed in Figure 2.21, the reduction of the degradation rate of organically modified matrices was accompanied by the preservation of the mesoporous arrangement of channels, as confirmed by the observed electron transmission micrographs after 60 days of essay.

As a consequence, the functionalization of the mesoporous channels not only allows the drug retention during molecular adsorption and modulation of the drug release kinetics to be increased, as mentioned previously, but also reduces the dissolution of the matrix. This effect is of particular interest because the release of the drugs would follow only one mechanism, the diffusion throughout the mesopore channels, rather than the additional erosion of the matrix.

The reduction of the degradation and control of the drug adsorption and release kinetics can also be achieved by the proper selection of the composition of the mesoporous carriers. In this sense, it is possible to stabilize silica mesoporous matrices by the addition of small amounts of zirconia during the synthetic process. The synthesis of these mixed oxides mesoporous materials can be easily performed through the EISA method using an aerosol route (Colilla et al. 2010). Figure 2.22 shows the electron microscopy evaluation performed after the loading and release processes, in which the mesostructure was not collapsed.

Additionally, the release of bisphosphonates from those matrices can be modulated by controlling the amount of zirconia within the mesoporous matrix. This effect takes place because of the complexation of the phosphonates from the drug

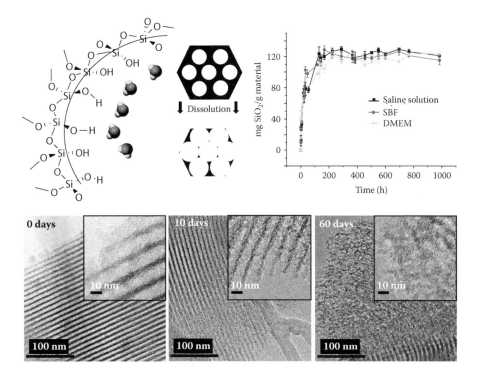

FIGURE 2.20 Graphical representation of the interaction of water with silica at the surface of the materials mesopores (top left corner), erosion of the mesoporous matrix (top center), silica concentration on different solutions versus time as a consequence of matrix dissolution (top right), and transmission electron micrographs of SBA-15 at different periods of time after soaking in aqueous media to study the degradation process of the matrix.

with the zirconia present in the matrix, which promotes drug retention, leading to a higher uptake and a more sustained release.

2.3 ORAL DRUG DELIVERY

Traditionally, amorphous silica has been generally regarded as safe and has been employed by pharmaceutical companies as an excipient for oral drug administration in the past (Moulari et al. 2008). In fact, amorphous silica was proposed as a drug delivery carrier back in the 1980s (Unger et al. 1983), and since then many different amorphous silica have been proposed as oral drug delivery systems (Simovic et al. 2011). Attending to the atomic order, mesoporous silica can be considered as amorphous, and these carriers are very attractive candidates to deliver drugs in which direct oral administration presents different drawbacks, such as poor aqueous solubility (Rigby, Fairhead, and Van der Walle 2008). This is the case with itraconazole, which is a triazole antifungal agent that is prescribed to patients with fungal infections. This drug presents a poor solubility in physiological environments, which leads to insufficient dissolution through the gastrointestinal tract. To improve this

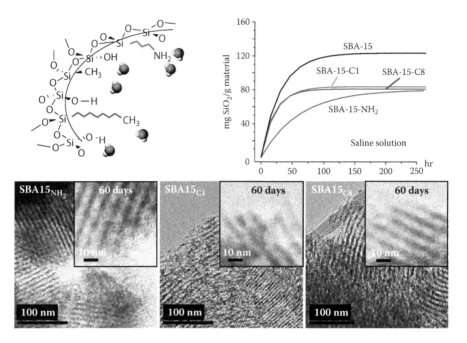

FIGURE 2.21 Graphical representation of the interaction of water with functionalized matrices at the surface of the material mesopores (top left corner), silica concentration on different solutions versus time as a consequence of organically modified matrix dissolution (top right), and transmission electron micrographs of SBA-15 modified with different organic groups after 60 days soaking in aqueous media to study the degradation process of the matrix.

situation, mesoporous silica has been employed to increase the bioavailability when orally delivered (Mellaerts, Mols et al. 2008). The performance of ordered mesoporous materials as drug carriers has been investigated in vivo using rabbits and dog models, evidencing that mesoporous silica material is a promising carrier to achieve improved oral bioavailability for drugs with poor solubility in aqueous media. In fact, mesoporous silica matrices offer unprecedented possibilities for the development of formulations for oral therapy (Mellaerts, Jammaer et al. 2008).

The employment of ordered mesoporous carriers for oral drug delivery is of particular interest for those systems that are activated with a change in the pH of the surrounding environment, so the drugs can be released at the stomach, intestine, or even the colon. In this sense, SBA-15 materials have been prepared in the shape of tablets for oral administration with a pH-sensitive polymer such as hydroxypropyl methylcellulose phthalate, which would allow controlling the release of the drug where it is needed. In simulated gastric fluid at pH 1.2, the model drug was released from unmodified SBA-15 in only two hours. However, coating the surface with the previously mentioned phthalate, the majority of the model drug was retained to be then released at pH 7.4 in a simulated intestinal fluid (Xu et al. 2009). In this way, ordered mesoporous materials can be employed to go through the stomach, avoiding any leaching of the cargo and protecting it from the acidic conditions, to then release the drug at the intestine. When dealing with intestinal diseases, this is a very smart

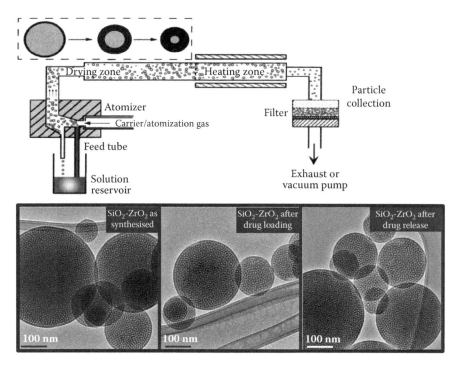

FIGURE 2.22 Scheme of the aerosol system employed for the synthesis of mixed oxide mesoporous materials (top) and transmission electron micrographs of SiO_2-ZrO_2 mesoporous particles as produced (bottom left), after zoledronate loading (bottom center) and after zoledronate release (bottom right).

approach to reach the intestine because in traditional medicine the ingested drug is adsorbed or digested by the stomach before reaching the intestine. Thus ordered mesoporous silica is a great carrier to be used as an intestinal biomolecule release system. A similar approach consists of coating amino-functionalized SBA-15 materials with poly(acrylic acid) to release a protein at the intestine. These materials were loaded with BSA as protein model, and it was observed that only 10% of the protein adsorbed was released after 36 hours at pH 1.2 (stomach-like). Thus the mesoporous carrier can protect proteins from being degraded from gastric enzymes, to then release those proteins at the intestine (Song, Hidajat, and Kawi 2007).

Although ordered mesoporous materials present a high pore volume and great surface area, which allow a great potential for molecular adsorption, the drug storage capacity can be increased by producing mesoporous silica spheres with pore channels penetrating from the outside to the inner hollow core. In this way, the hollow core can confine three times more drug molecules than MCM-41, which would be released following a diffusion mechanism, as in conventional mesoporous materials. Additionally, these systems can be easily coated with a polyelectrolyte multilayer, which can be designed to release the molecules at the desired pH. In this way, the drugs can be released at the stomach or at the intestine, as desired (Zhu et al. 2005).

Ordered mesoporous materials have found application in the treatment of ulcers in the stomach and intestine when employing them as an oral famotidine delivery system. Famotidine is a histamine H2-receptor antagonist with a half-life of elimination of only 2.6 hours. Unfortunately, the bioavailability of oral intakes reaches only 43% with severe fluctuations of concentration. To solve all these problems, MSU materials functionalized with COOH groups have been employed as oral delivery systems (Tang et al. 2006). However, the loading capacity of these particular mesoporous materials was not the best, so the same approach was carried out using SBA-15 materials bifunctionalized with COOH to favor the famotidine loading and trimethylsilyl to improve the release characteristics of the matrix carrier (Xu et al. 2008).

Ordered mesoporous materials can also be employed as colon-specific drug delivery systems when orally taken (S.-H. Cheng et al. 2011). In this sense, ordered mesoporous carriers can be designed to avoid the degradation and the nonspecific release of the cargo in the gastrointestinal tract under acidic conditions. To achieve this, the mesoporous silica surface can be modified with pH-sensitive trimethylammonium groups through a pH-sensitive hydrazine bond. After the oral administration of this carrier, the acidic pH of the stomach can fully hydrolyze the trimethylammonium-hydrazone bonds. Then the drug molecules can be released where the environmental pH is close to neutral, such as the colon. Thus mesoporous silica can be employed as oral drug delivery systems for colon-related diseases with improved site specificity and release kinetics of different pharmaceutical agents.

Recently, ordered mesoporous materials have been proposed for the oral delivery of anticancer drugs to block Notch signaling in the intestine, which is frequently activated in cancer (Mamaeva et al. 2011). In vitro testing induced cell-specific inhibition of Notch activity and exhibited enhanced tumor retainment. The oral administration of these carriers controlled Notch activity in intestinal stem cells, and after systemic administration the particles showed tumor accumulation and targeting. For this reason, ordered mesoporous materials are very promising materials to be employed in the treatment of certain cancer diseases with restricted clinical application, such as intestinal cancer.

2.4 TOPICAL AND TRANSDERMAL DRUG DELIVERY

Although the terms *topical* and *transdermal* are sometimes interchanged, the concepts are different. A topical medication is designed to have an effect at the site of application to treat normally dermal disorders, while transdermal medications, which are absorbed through the skin or mucosal membranes, are designed to have an effect in different areas of the living body away from the site of application. Thus topical drug delivery can be defined as the application of a drug-containing formulation to the skin through the employment of foams, spray, medicated powders, solution, and even medicated adhesive systems. The drugs that are normally released from topical delivery systems are antibiotics to treat skin infections, corticosteroids for skin irritation, and some topical anesthetics. The main advantage of this technology is that the drug is delivered where it is needed while minimizing it at places where it is not needed or wanted. Regarding the use of ordered mesoporous materials for topical delivery, few reports have been published, since the medications are normally dispensed in skin-friendly agents, such as ointments, creams, gels, or pastes. Mesoporous silica nanocomposites as topical drug delivery systems have been investigated with caffeine as the model drug, and in vitro caffeine diffusion into and through the skin has been evaluated in comparison with a caffeine gel formulation using newborn pig skin and vertical Franz diffusion cells (Pilloni et al. 2012). However, this is just a basic research work, with no further in vivo model investigation.

On the other hand, transdermal delivery systems are placed on the skin to deliver a specific dose of medication through the skin into the bloodstream. This type of technology is applied when the patient is unable to swallow or for medications that are significantly metabolized by the liver. The main advantage of the transdermal delivery system is that it can provide a controlled release of the medication into the patient. Ordered mesoporous materials can be of great applicability since they have been shown to modulate the release rate of the adsorbed drugs. These materials can act as drug reservoirs, and when used as additives of certain polymers they can be part of adhesive patches and modulate the release of the drugs throughout the diffusion from the mesopores.

2.5 TARGETED DRUG DELIVERY

In the last few years, targeted drug delivery therapies are acquiring a great relevance within biomedical sciences because of the potential effectiveness that they can offer. The basis of targeted systems is the fact that biomolecular pairs exhibit specificity and high affinity for each other within living bodies. The application of this idea to drug delivery systems results in targeted drug delivery technology. One of the biomolecular entities is attached to the surface of the carrier system that can therefore specifically bind to the other biomolecular entity at the target site.

Targeted drug delivery systems are promising tools for the treatment of certain diseases such as cancer, in which traditional therapies rely on nonspecific action that presents serious side effects on healthy tissues. The use of targeted drug delivery systems would permit a better efficacy, thanks to the controlled delivery from the carrier matrices, and specific destruction of cancer cells, thanks to selective targeting to the surface receptors overexpressed on tumor cells. In this sense, the key aspect of this technology is the capability of specific ligands or antibodies to selectively bind tumor cells through molecular recognition (Allen 2002).

Most of the targeted delivery systems employing ordered mesoporous materials have been developed using mesoporous silica nanoparticles as delivery platforms. The scientific interest and number of publications in the last few years on targeted delivery systems is so high that the authors of this book had considered devoting a whole chapter just to this topic. In this sense, Chapter 4 of this book will be dedicated to those mesoporous nanoparticles in biomedicine, with special attention to targeted delivery systems for cancer therapy, where promising results will be detailed.

REFERENCES

Allen, T. M. 2002. "Ligand-Targeted Therapeutics in Anticancer Therapy." *Nature Reviews Cancer* 2 (10):750–63.

Allouche, J., M. Boissiere, C. Helary, J. Livage, and T. Coradin. 2006. "Biomimetic Core-Shell Gelatine/Silica Nanoparticles: A New Example of Biopolymer-Based Nanocomposites." *Journal of Materials Chemistry* 16 (30):3120–25.

Areva, S., V. Aaritalo, S. Tuusa, M. Jokinen, M. Linden, and T. Peltola. 2007. "Sol-Gel-Derived TiO2-SiO2 Implant Coatings for Direct Tissue Attachment. Part II: Evaluation of Cell Response." *Journal of Materials Science: Materials in Medicine* 18 (8):1633–42.

Arruebo, M., N. Vilaboa, and J. Santamaria. 2010. "Drug Delivery from Internally Implanted Biomedical Devices Used in Traumatology and in Orthopedic Surgery." *Expert Opinion on Drug Delivery* 7 (5):589–603.

Balas, F., M. Manzano, M. Colilla, and M. Vallet-Regí. 2008. "L-Trp Adsorption into Silica Mesoporous Materials to Promote Bone Formation." *Acta Biomaterialia* 4 (3):514–22.

Balas, F., M. Manzano, P. Horcajada, and M. Vallet-Regí. 2006. "Confinement and Controlled Release of Bisphosphonates on Ordered Mesoporous Silica-Based Materials." *Journal of the American Chemical Society* 128 (25):8116–17.

Bass, J. D., D. Grosso, C. Boissiere, E. Belamie, T. Coradin, and C. Sanchez. 2007. "Stability of Mesoporous Oxide and Mixed Metal Oxide Materials under Biologically Relevant Conditions." *Chemistry of Materials* 19 (17):4349–56.

Blin, J. L., C. Otjacques, G. Herrier, and B. L. Su. 2000. "Pore Size Engineering of Mesoporous Silicas Using Decane as Expander." *Langmuir* 16 (9):4229–36.

Bottini, M., F. D'Annibale, A. Magrini, F. Cerignoli, Y. Arimura, M. I. Dawson, E. Bergamaschi, N. Rosato, A. Bergamaschi, and T. Mustelin. 2007. "Quantum Dot-Doped Silica Nanoparticles as Probes for Targeting of T-lymphocytes." *International Journal of Nanomedicine* 2 (2):227–33.

Brunauer, S., L. S. Deming, W. E. Deming, and E. Teller. 1940. "On a Theory of the Van der Waals Adsorption of Gases." *Journal of the American Chemical Society* 62:1723–32.

Charnay, C., S. Begu, C. Tourne-Peteilh, L. Nicole, D. A. Lerner, and J. M. Devoisselle. 2004. "Inclusion of Ibuprofen in Mesoporous Templated Silica: Drug Loading and Release Property." *European Journal of Pharmaceutics and Biopharmaceutics* 57 (3):533–40.

Cheng, K., and C. C. Landry. 2007. "Diffusion-Based Deprotection in Mesoporous Materials: A Strategy for Differential Functionalization of Porous Silica Particles." *Journal of the American Chemical Society* 129:9674–85.

Cheng, Shih-Hsun, Wei-Neng Liao, Li-Ming Chen, and Chia-Hung Lee. 2011. "pH-Controllable Release Using Functionalized Mesoporous Silica Nanoparticles as an Oral Drug Delivery System." *Journal of Materials Chemistry* 21 (20):7130–37.

Colilla, M., M. Manzano, I. Izquierdo-Barba, M. Vallet-Regí, C. Boissiere, and C. Sanchez. 2010. "Advanced Drug Delivery Vectors with Tailored Surface Properties Made of Mesoporous Binary Oxides Submicronic Spheres." *Chemistry of Materials* 22 (5):1821–30.

Doadrio, A. L., E. M. B. Sousa, J. C. Doadrio, J. P. Pariente, I. Izquierdo-Barba, and M. Vallet-Regí. 2004. "Mesoporous SBA-15 HPLC Evaluation for Controlled Gentamicin Drug Delivery." *Journal of Controlled Release* 97 (1):125–32.

Dormer, K., C. Seeney, K. Lewelling, G. D. Lian, D. Gibson, and M. Johnson. 2005. "Epithelial Internalization of Superparamagnetic Nanoparticles and Response to External Magnetic Field." *Biomaterials* 26 (14):2061–72.

Gartmann, N., and D. Bruhwiler. 2009. "Controlling and Imaging the Functional-Group Distribution on Mesoporous Silica." *Angewandte Chemie—International Edition* 48 (34):6354–56.

Gerion, D., J. Herberg, R. Bok, E. Gjersing, E. Ramon, R. Maxwell, J. Kurhanewicz, T. F. Budinger, J. W. Gray, M. A. Shuman, and F. F. Chen. 2007. "Paramagnetic Silica-Coated Nanocrystals as an Advanced MRI Contrast Agent." *Journal of Physical Chemistry C* 111 (34):12542–51.

Gonzalez, B., M. Colilla, C. L. de Laorden, and M. Vallet-Regí. 2009. "A Novel Synthetic Strategy for Covalently Bonding Dendrimers to Ordered Mesoporous Silica: Potential Drug Delivery Applications." *Journal of Materials Chemistry* 19 (47):9012–24.

Horcajada, P., A. Ramila, F. Gerard, and M. Vallet-Regí. 2006. "Influence of Superficial Organic Modification of MCM-41 Matrices on Drug Delivery Rate." *Solid State Sciences* 8 (10):1243–49.

Horcajada, P., A. Ramila, J. Perez-Pariente, and M. Vallet-Regí. 2004. "Influence of Pore Size of MCM-41 Matrices on Drug Delivery Rate." *Microporous and Mesoporous Materials* 68 (1–3):105–9.

Huo, Q. S., J. Liu, L. Q. Wang, Y. B. Jiang, T. N. Lambert, and E. Fang. 2006. "A New Class of Silica Cross-Linked Micellar Core-Shell Nanoparticles." *Journal of the American Chemical Society* 128 (19):6447–53.

Izquierdo-Barba, I., M. Colilla, M. Manzano, and M. Vallet-Regí. 2010. "In Vitro Stability of SBA-15 under Physiological Conditions." *Microporous and Mesoporous Materials* 132 (3):442–52.

Kecht, J., A. Schlossbauer, and T. Bein. 2008. "Selective Functionalization of the Outer and Inner Surfaces in Mesoporous Silica Nanoparticles." *Chemistry of Materials* 20 (23):7207–14.

Kimura, T., Y. Sugahara, and K. Kuroda. 1998. "Synthesis of Mesoporous Aluminophosphates Using Surfactants with Long Alkyl Chain Lengths and Triisopropylbenzene as a Solubilizing Agent." *Chemical Communications* (5):559–60.

Kohane, D. S., and R. Langer. 2010. "Biocompatibility and Drug Delivery Systems." *Chemical Science* 1 (4):441–46.

Kokubo, T., and H. Takadama. 2006. "How Useful Is SBF in Predicting In Vivo Bone Bioactivity?" *Biomaterials* 27 (15):2907–15.

Kruk, M., M. Jaroniec, and A. Sayari. 2000. "New Insights into Pore-Size Expansion of Mesoporous Silicates Using Long-Chain Amines." *Microporous and Mesoporous Materials* 35–36:545–53.

Lei, Jie, Jie Fan, Chengzhong Yu, Luyan Zhang, Shiyi Jiang, Bo Tu, and Dongyuan Zhao. 2004. "Immobilization of Enzymes in Mesoporous Materials: Controlling the Entrance to Nanospace." *Microporous and Mesoporous Materials* 73 (3):121–28.

Lettow, J. S., Y. J. Han, P. Schmidt-Winkel, P. D. Yang, D. Y. Zhao, G. D. Stucky, and J. Y. Ying. 2000. "Hexagonal to Mesocellular Foam Phase Transition in Polymer-Templated Mesoporous Silicas." *Langmuir* 16 (22):8291–95.

Lopez-Noriega, A., D. Arcos, and M. Vallet-Regí. 2010. "Functionalizing Mesoporous Bioglasses for Long-Term Anti-osteoporotic Drug Delivery." *Chemistry—a European Journal* 16 (35):10879–86.

Mamaeva, Veronika, Jessica M. Rosenholm, Laurel Tabe Bate-Eya, Lotta Bergman, Emilia Peuhu, Alain Duchanoy, Lina E. Fortelius, Sebastian Landor, Diana M. Toivola, Mika Linden, and Cecilia Sahlgren. 2011. "Mesoporous Silica Nanoparticles as Drug Delivery Systems for Targeted Inhibition of Notch Signaling in Cancer." *Molecular Therapy* 19 (8):1538–46.

Manzano, M., G. Lamberti, I. Galdi, and M. Vallet-Regí. 2011. "Anti-osteoporotic Drug Release from Ordered Mesoporous Bioceramics: Experiments and Modeling." *Aaps Pharmscitech* 12 (4):1193–99.

Manzano, M., and M. Vallet-Regí. 2010. "New Developments in Ordered Mesoporous Materials for Drug Delivery." *Journal of Materials Chemistry* 20 (27):5593–5604.

Mellaerts, R., J. A. G. Jammaer, M. van Speybroeck, H. Chen, J. van Humbeeck, P. Augustijns, G. van den Mooter, and J. A. Martens. 2008. "Physical State of Poorly Water Soluble Therapeutic Molecules Loaded into SBA-15 Ordered Mesoporous Silica Carriers: A Case Study with Itraconazole and Ibuprofen." *Langmuir* 24 (16):8651–59.

Mellaerts, R., R. Mols, J. A. G. Jammaer, C. A. Aerts, P. Annaert, J. van Humbeeck, G. van den Mooter, P. Augustijns, and J. A. Martens. 2008. "Increasing the Oral Bioavailability of the Poorly Water Soluble Drug Itraconazole with Ordered Mesoporous Silica." *European Journal of Pharmaceutics and Biopharmaceutics* 69 (1):223–30.

Menaa, B., F. Menaa, C. Aiolfi-Guimaraes, and O. Sharts. 2010. "Silica-Based Nanoporous Sol-Gel Glasses: From Bioencapsulation to Protein Folding Studies." *International Journal of Nanotechnology* 7 (1):1–45.

Menaa, B., Y. Miyagawa, M. Takahashi, M. Herrero, V. Rives, F. Menaa, and D. K. Eggers. 2009. "Bioencapsulation of Apomyoglobin in Nanoporous Organosilica Sol-Gel Glasses: Influence of the Siloxane Network on the Conformation and Stability of a Model Protein." *Biopolymers* 91 (11):895–906.

Moulari, B., D. Pertuit, Y. Pellequer, and A. Lamprecht. 2008. "The Targeting of Surface Modified Silica Nanoparticles to Inflamed Tissue in Experimental Colitis." *Biomaterials* 29 (34):4554–60.

Nieto, A., F. Balas, M. Colilla, M. Manzano, and M. Vallet-Regí. 2008. "Functionalization Degree of SBA-15 as Key Factor to Modulate Sodium Alendronate Dosage." *Microporous and Mesoporous Materials* 116 (1–3):4–13.

Pilloni, Martina, Guido Ennas, Mariano Casu, Anna Maria Fadda, Francesca Frongia, Francesca Marongiu, Roberta Sanna, Alessandra Scano, Donatella Valenti, and Chiara Sinico. 2012. "Drug Silica Nanocomposite: Preparation, Characterization and Skin Permeation Studies." *Pharmaceutical Development and Technology* (February 11):1–8.

Qu, F. Y., G. S. Zhu, S. Y. Huang, S. G. Li, J. Y. Sun, D. L. Zhang, and S. L. Qiu. 2006. "Controlled Release of Captopril by Regulating the Pore Size and Morphology of Ordered Mesoporous Silica." *MicroporouRigby, S. P., M. Fairhead, and C. F. van der Walle. 2008. "Engineering Silica Particles as Oral Drug Delivery Vehicles." Current Pharmaceutical Design* 14 (18):1821–31.

Sanchez, C., C. Boissiere, D. Grosso, C. Laberty, and L. Nicole. 2008. "Design, Synthesis, and Properties of Inorganic and Hybrid Thin Films Having Periodically Organized Nanoporosity." *Chemistry of Materials* 20 (3):682–737.

Schmidt-Winkel, P., W. W. Lukens, D. Y. Zhao, P. D. Yang, B. F. Chmelka, and G. D. Stucky. 1999. "Mesocellular Siliceous Foams with Uniformly Sized Cells and Windows." *Journal of the American Chemical Society* 121 (1):254–55.

Shephard, D. S., W. Z. Zhou, T. Maschmeyer, J. M. Matters, C. L. Roper, S. Parsons, B. F. G. Johnson, and M. J. Duer. 1998. "Site-Directed Surface Derivatization of MCM-41: Use of High-Resolution Transmission Electron Microscopy and Molecular Recognition for Determining the Position of Functionality within Mesoporous Materials." *Angewandte Chemie—International Edition* 37 (19):2719–23.

Simovic, Spomenka, Nasrin Ghouchi-Eskandar, Aw Moom Sinn, Dusan Losic, and Clive A. Prestidge. 2011. "Silica Materials in Drug Delivery Applications." *Current Drug Discovery Technologies* 8 (3):269–76.

Slowing, II, J. L. Vivero-Escoto, C. W. Wu, and V. S. Y. Lin. 2008. "Mesoporous Silica Nanoparticles as Controlled Release Drug Delivery and Gene Transfection Carriers." *Advanced Drug Delivery Reviews* 60 (11):1278–88.

Smarsly, B., S. Polarz, and M. Antonietti. 2001. "Preparation of Porous Silica Materials via Sol-Gel Nanocasting of Nonionic Surfactants: A Mechanistic Study on the Self-Aggregation of Amphiphiles for the Precise Prediction of the Mesopore Size." *Journal of Physical Chemistry B* 105 (43):10473–83.

Song, S. W., K. Hidajat, and S. Kawi. 2007. "pH-Controllable Drug Release Using Hydrogel Encapsulated Mesoporous Silica." *Chemical Communications* (42):4396–98.

Tang, Q. L., Y. Xu, D. Wu, and Y. H. Sun. 2006. "A Study of Carboxylic-Modified I-Mesoporous Silica in Controlled Delivery for Drug Famotidine." *Journal of Solid State Chemistry* 179 (5):1513–20.

Unger, K., H. Rupprecht, B. Valentin, and W. Kircher. 1983. "The Use of Porous and Surface Modified Silicas as Drug Delivery and Stabilizing Agents." *Drug Development and Industrial Pharmacy* 9 (1–2):69–91.

Vallet-Regí, M., F. Balas, M. Colilla, and M. Manzano. 2008. "Bone-Regenerative Bioceramic Implants with Drug and Protein Controlled Delivery Capability." *Progress in Solid State Chemistry* 36 (3):163–91.

Vallet-Regí, M., M. Manzano, J. M. Gonzalez-Calbet, and E. Okunishi. 2010. "Evidence of Drug Confinement into Silica Mesoporous Matrices by STEM Spherical Aberration Corrected Microscopy." *Chemical Communications* 46 (17):2956–58.

Vallet-Regí, M., A. Ramila, R. P. del Real, and J. Perez-Pariente. 2001. "A New Property of MCM-41: Drug Delivery System." *Chemistry of Materials* 13 (2):308–11.

Xu, W. J., Q. Gao, Y. Xu, D. Wu, and Y. H. Sun. 2009. "pH-Controlled Drug Release from Mesoporous Silica Tablets Coated with Hydroxypropyl Methylcellulose Phthalate." *Materials Research Bulletin* 44 (3):606–12.

Xu, Wujun, Qiang Gao, Yao Xu, Dong Wu, Yuhan Sun, Wanling Shen, and Feng Deng. 2008. "Controlled Drug Release from Bifunctionalized Mesoporous Silica." *Journal of Solid State Chemistry* 181 (10):2837–44.

Yiu, H. H. P., and P. A. Wright. 2005. "Enzymes Supported on Ordered Mesoporous Solids: A Special Case of an Inorganic-Organic Hybrid." *Journal of Materials Chemistry* 15 (35–36):3690–3700.

Yiu, H. H. P., P. A. Wright, and N. P. Botting. 2001. "Enzyme Immobilisation using SBA-15 Mesoporous Molecular Sieves with Functionalised Surfaces." *Journal of Molecular Catalysis B: Enzymatic* 15 (1–3):81–92.

Zhao, Jianwei, Feng Gao, Yunlin Fu, Wan Jin, Pengyuan Yang, and Dongyuan Zhao. 2002. "Biomolecule Separation Using Large Pore Mesoporous SBA-15 as a Substrate in High Performance Liquid Chromatography." *Chemical Communications* (7):752–53.

Zhu, Y. F., J. L. Shi, W. H. Shen, X. P. Dong, J. W. Feng, M. L. Ruan, and Y. S. Li. 2005. "Stimuli-Responsive Controlled Drug Release from a Hollow Mesoporous Silica Sphere/Polyelectrolyte Multilayer Core-Shell Structure." *Angewandte Chemie-International Edition* 44 (32):5083–87.

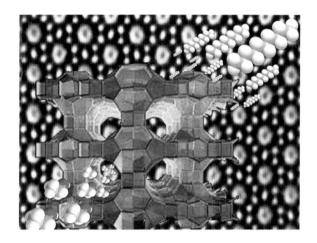

FIGURE 1 Schematic representation of the ZSM-5 catalyst, which is made of silica and alumina and was employed to transform alcohol into gasoline.

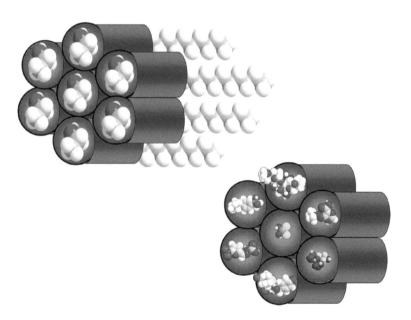

FIGURE 2 Schematic representations of the conversion of alcohol in gasoline (top) and confinement of different pharmaceutical agents into the mesopores of the silica matrices.

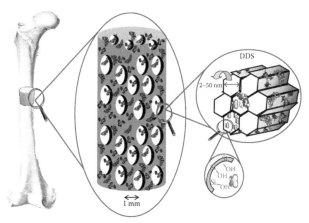

FIGURE 1.2 Scheme of a potential scaffold made of mesoporous bioceramics and with bone regenerative character together with drug delivery capabilities.

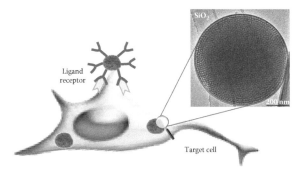

FIGURE 1.17 Schematic representation of mesoporous nanoparticles with targeting agents grafted on their surface to specifically interact with certain receptors to be then internalized to release their cargo.

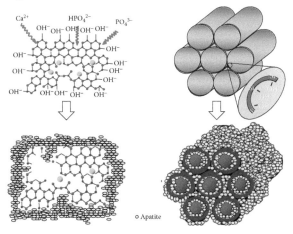

FIGURE 1.27 Representation of the formation of a new apatite-like layer on the surface of glasses (left) and ordered mesoporous materials (right) when in contact with physiological fluids.

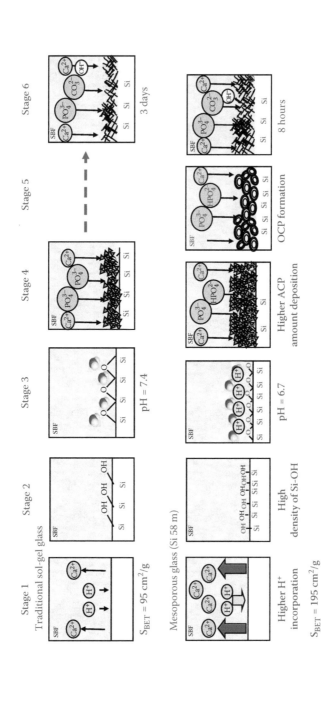

FIGURE 1.38 Bioactive mechanism in SBF of traditional sol-gel glasses (proposed by Hench in 1970) (Hench 1991) compared to bioactive mechanism of MBGs proposed by Izquierdo-Barba et al. (2008).

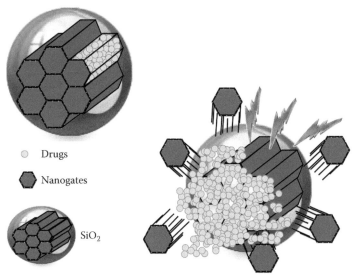

FIGURE 3.1 Schematic representation of stimuli-responsive drug delivery from silica-based ordered mesoporous materials. The different stimuli that have been used to trigger drug delivery include pH, temperature, redox potential, light, magnetic field, ultrasounds, enzymes, antibodies, aptamer targets, or even the combination of more than one stimulus.

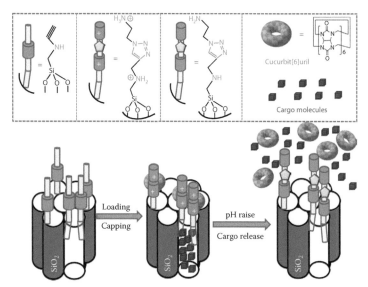

FIGURE 3.3 Graphical representation of the performance of pH-responsive supramolecular valves based in [2]pseudorotaxanes consisting of bisammonium stalks and cucurbit[6] uril (CB[6]) rings. The alkyne-functionalized MCM-41 is loaded with a model molecule and capped with CB[6] during the CB[6]-catalyzed alkyne-azide 1,3-dipolar cycloadditions, followed by washing away the excess of substrates. Entrapped cargo is released by switching off the ion-dipole interactions between the CB[6] rings and the bisammonium stalks upon raising the pH value. *Source*: Angelos et al. (2008).

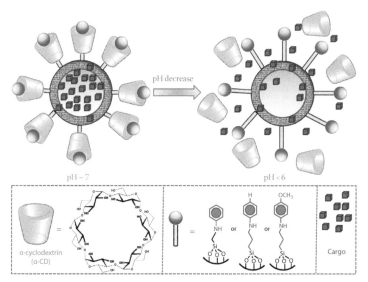

FIGURE 3.5 Schematic representation of pH-responsive operating mechanism of hollow MSNPs capped by supramolecular machines based on α-CD. At pH ~7, the aniline nitrogens on the stalks are not protonated and are encircled by the α-CD rings that block the pores. When the pH is decreased, protonation of the nitrogen causes α-CD to dissociate, unblocking the pores and releasing the trapped cargo molecules. *Source*: Du et al. 2009).

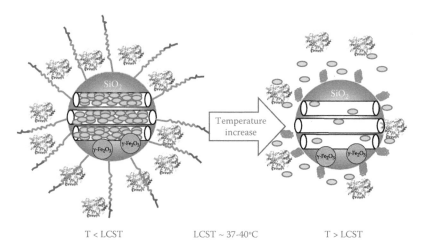

FIGURE 3.8 Schematic depiction of the temperature-responsive performance of MSNPs encapsulating iron oxide nanocrystals and decorated on the surface with a thermo-responsive copolymer of poly(ethyleneimine)-b-poly(N-isopropylacrylamide) (PEI/NIPAM). The polymer acts both as temperature-responsive gatekeeper for the drugs trapped inside the silica matrix and retains proteins in the polymer shell by electrostatic or hydrogen bond interactions. The nanocarrier traps the different cargos at low temperatures (20°C) and releases the retained molecules when the temperature exceeds 35–40°C following different kinetics. *Source*: Baeza et al. (2012).

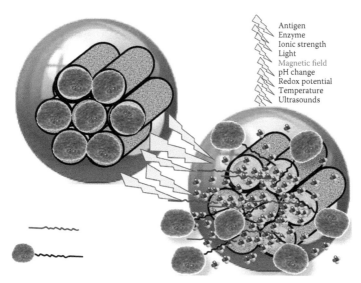

Antigen
Enzyme
Ionic strength
Light
Magnetic field
pH change
Redox potential
Temperature
Ultrasounds

FIGURE 3.14 Progressive double-stranded DNA dehybridization upon the application of an appropriate stimulus. In this case, the removal of the DNA/magnetic nanocap conjugates and the subsequent cargo release take place as a result of temperature increase due to the application of an alternating magnetic field. *Source*: Ruiz-Hernández, Baeza, and Vallet-Regí (2011).

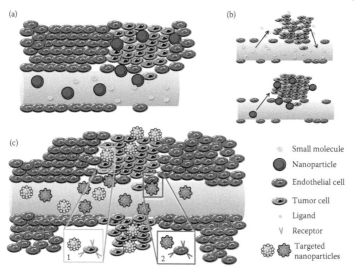

(a) (b) (c)

Small molecule
Nanoparticle
Endothelial cell
Tumor cell
Ligand
Receptor
Targeted nanoparticles

FIGURE 4.2 (a) Passive targeting of nanoparticles, which reach tumor cells selectively through the leaky vasculature surrounding the tumors. (b) Influence of the size in the retention in the tumor tissue. Small molecules, such as drugs, diffuse without restraint in and out of the tumor blood vessels, which prevents their effective accumulation in the tumor. By contrast, the relatively large size of nanoparticles (20–200 nm) prevents their diffusion back into the blood stream, which results in progressive accumulation, i.e., the EPR effect. (c) Strategies for active targeting of nanoparticles. Ligands anchored on the surface of nanoparticles bind to receptors overexpressed by (1) cancer cells or (2) angiogenic endothelial cells. *Sources*: Muggia (1999); Maeda, Bharate, and Daruwalla (2009); Torchilin (2011); Langer (1998).

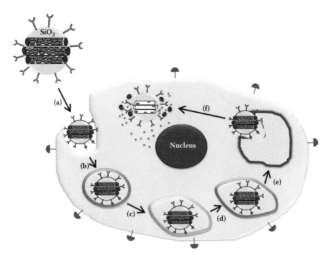

FIGURE 4.3 Schematic representation of the intracellular trafficking of MSNPs. (a) The cellular uptake takes place through specific (ligand-receptor) and nonspecific (hydrophobic, Coulombic) binding interactions. (b) The internalized MSNPs are delivered to intermediate compartments, such as caveosomes. (c) Such compartments are transported to early endosomes and then to sorting endosomes. From sorting endosomes a fraction of the MSNPs are sorted back to the cell exterior through recycling endosomes (not shown in this figure. (d) The remaining fraction is transported to secondary endosomes, (e) which then fuse with lysosomes. (f) The MSNs escape the endolysosomes and enter the cytosolic compartment. Then the MSNPs can release their cargo under a given change in the surrounding medium, which acts as a stimulus that triggers drug release.

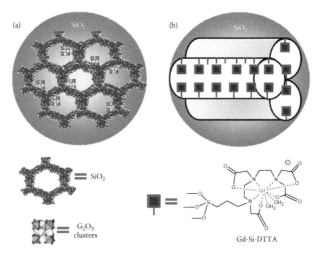

FIGURE 4.7 Schematic representations of MSNPs designed as highly efficient MRI contrast agents by: (a) incorporating Gd_2O_3 clusters on the inner surface of mesopores (Li, Liu et al. 2011) or (b) functionalizing silica walls with Gd-Si-DTTA complex (Si-DTTA = 3-aminopropyl(trimethoxysilyl) diethylenetriamine tetraacetic acid) via siloxane linkage. *Source*: Taylor et al. (2008).

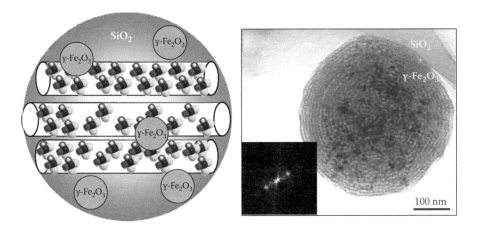

FIGURE 4.9 Schematic representation of MSNPs encapsulating γ-Fe$_2$O$_3$ nanoparticles and loaded with drugs. Transmission electron microscopy (TEM) image of the system and its corresponding Fourier transform are also displayed.

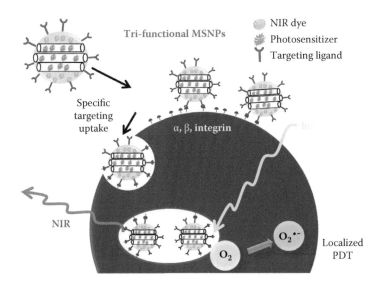

FIGURE 4.12 Schematic representation of tri-functional MSNPs as theranostic agents. The presence of RGD peptides grafted to the external surface of MSNPs permits targeting the overexpressed $\alpha_v\beta_3$ integrins of cancer cells. The NIR dye (ATTO647N) incorporated within the silica network allows traceable imaging of particles. The photosensitizer *meso*-tetratolylporphyrin-Pd (PdTPP) grafted to the mesopores enables local PDT. *Source*: Cheng et al. (2010).

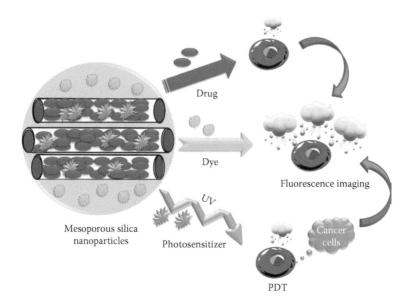

FIGURE 4.15 MSNPs for theranostic applications, combining fluorescence imaging (by incorporation of a fluorescent dye) and dual-mode cancer therapy (chemotherapy and photo-dynamic therapy [PDT]).The synergy of combining chemotherapy and PDT exhibiting high therapeutic efficacy for cancer cells is also schematically displayed.

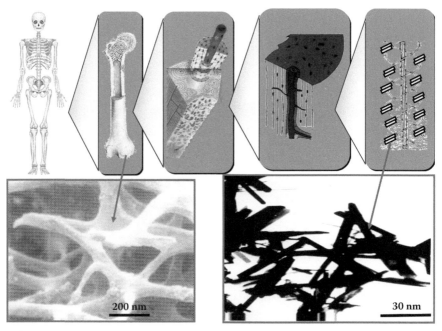

FIGURE 5.3 Representation of the hierarchical structure of bone (top); and micrographs of bone macroporosity (bottom left) and carbonate hydroxyapatite (bottom right).

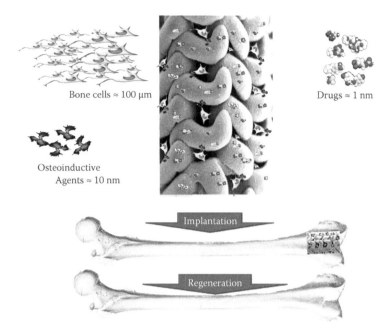

FIGURE 5.7 Schematic representation of scaffolds with osteoinductive agents on their surface that promote cell attachment and proliferation, and also are able to release certain drugs for diverse bone pathological situations.

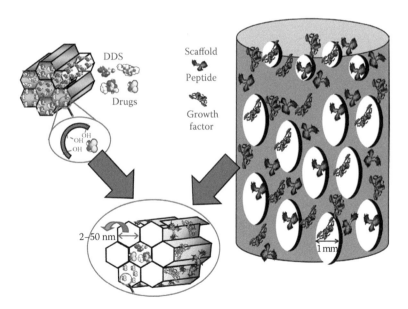

FIGURE 5.11 Future scaffolds made of ordered mesoporous materials with two functions: bone tissue regeneration and drug delivery capabilities.

3 Stimuli-Responsive Drug Delivery Systems Based on Mesoporous Silica

The control over the release of molecules from mesoporous materials is a pivotal factor in developing efficient drug delivery devices. Conventional drug delivery systems, which have been described in Chapter 2, involve the release of adsorbed molecules following sustained release mechanisms that can be expressed in terms of diffusion of adsorbed molecules throughout the mesopore channels in the silica matrix. Consequently, release kinetics can be understood in terms of the Fickean diffusion coefficients, which depend on the characteristics of both the guest molecule and the host matrix. However, certain biomedical applications require the development of stimuli-responsive release systems (Shi et al. 2010). For instance, many site-selective delivery systems, such as those for highly toxic antitumor drugs, require zero release before reaching the targeted cell or tissue. In this sense, it is possible to use the available channels of mesoporous materials as drug reservoirs and block the pore entrances with appropriate nanogates. For example, inorganic nanoparticles, polymers, and larger supramolecular assemblies have been used as the blocking caps to control opening/closing of pore entrances of mesoporous silica. Different stimuli, pH, temperature, redox potential, light, magnetic field, ultrasounds, enzymes, antibodies, aptamer targets, or even the combination of more than one stimulus have been used as triggers for uncapping the pores and releasing the guest molecules from mesoporous silica (Figure 3.1) (Ambrogio et al. 2011; Coti et al. 2009; Liong et al. 2008, 2009; Manzano and Vallet-Regí 2010; Saha et al. 2007; Slowing et al. 2008; Trewyn et al. 2007; Vallet-Regí et al. 2011; Vallet-Regí, Balas, and Arcos 2007; Ashley et al. 2011; R. Liu, Zhang, and Feng 2009; Vallet-Regí, Colilla, and Gonzalez 2011; Vallet-Regí and Ruiz-Hernández 2011; Manzano, Colilla, and Vallet-Regí 2009; J. Liu et al. 2009).

To date, most of the research groups have focused their scientific interest and achievements on stimuli-responsive drug delivery from mesoporous silica nanoparticles (MSNPs). However, the acquired knowledge can be moved into the bulk dimension with the aim of developing silica mesoporous scaffolds (Figure 3.2). This would permit the design of a smart and multifunctional scaffold with potential applications in the bone tissue regeneration field.

This chapter illustrates the different approaches developed to date aimed at designing stimuli-responsive drug release mesoporous silica systems. To give a clear overview of the current state of the art in this mushrooming research field, the

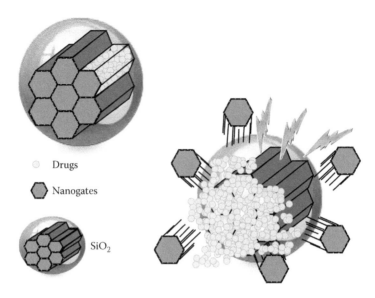

FIGURE 3.1   (See color insert.) Schematic representation of stimuli-responsive drug delivery from silica-based ordered mesoporous materials. The different stimuli that have been used to trigger drug delivery include pH, temperature, redox potential, light, magnetic field, ultrasounds, enzymes, antibodies, aptamer targets, or even the combination of more than one stimulus.

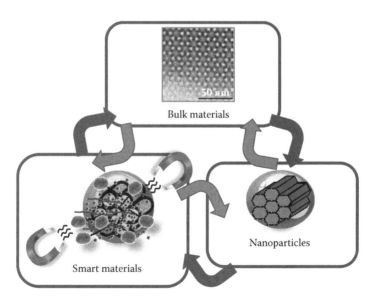

FIGURE 3.2   Scheme of the scientific knowledge directions in the past, present, and future.

description of the design and operation mechanisms of the different stimuli-responsive delivery systems focuses on the stimulus used as release trigger.

3.1   pH AS RELEASE TRIGGER

pH is an attractive release trigger since some tissues in the body, such as tumors, inflamed tissues, as well as endosomal cell compartments, have a more acidic pH than blood or healthy tissues.

Zink's group has wide research experience in developing pH-responsive mechanized MSNPs by using mechanically interlocked macromolecules (Ambrogio et al. 2011; Coti et al. 2009; Saha et al. 2007). For this purpose, they have used cyclodextrins (CDs) and cucurbit[6]uril (CB[6]), since both polymacrocycles tend to form inclusion complexes with a variety of guest molecules in aqueous solution, which can be dissociated reversibly in response to changes in the environmental pH. In their early investigations they developed pH-responsive MSNPs based on the ion-dipole interaction between CB[6] and bisammonium stalks, which operated in water (Angelos et al. 2008). CB[6] is a pumpkin-shaped polymacrocycle consisting of six glycouril units strapped together by pairs of bridging methylene groups between nitrogen atoms (Day et al. 2001; K. Kim 2002; Lagona et al. 2005; Lee et al. 2003; S. Liu et al. 2005). CB[6] has the ability to form inclusion complexes with a variety of polymethylene derivatives, especially diamino-alkanes. The stabilities of these 1:1 complexes are highly pH dependent, and therefore CB[6] can be used as a gatekeeper to design pH-responsive mesoporous silica release systems. Thus [2]pseudorotaxanes consisting of bisammonium stalks and CB[6] rings were prepared on the surface of MSNPs, and the pH-dependent association/dissociation of the CB[6] ring with diaminoalkanes enabled the formation of complexes whose dynamic behavior can be controlled by pH (Figure 3.3) (Angelos et al. 2008). Thus at neutral and acidic pH values, the CB[6] rings encircle the bisammonium stalks tightly, therefore blocking the mesopore outlets efficiently when using tethers of suitable lengths and preventing the departure of the entrapped cargo. Deprotonation of the stalks after addition of base results in dethreading of the CB[6] rings and unblocking of the mesopores, which allows the release of the entrapped molecules.

However, with the aim of designing tunable pH-operable MSNPs in which biologically relevant pH variations are used to trigger the release of entrapped guest molecules, Zink and coworkers redesigned the CB[6] pseudorotaxane functionalized MSNPs to remain closed at neutral pH (i.e., the bloodstream) but open under mildly acidic conditions, such as the lysosomes, releasing their contents autonomously upon cell uptake (Angelos, Khashab et al. 2009). For this purpose they used trisammonium stalks that contain one anilinium and two $-CH_2NH_2^+CH_2$ centers (Figure 3.4). Since the anilinium nitrogen atom is ca. 10^6-fold less basic than the alkyl nitrogen atoms, it remains not protonated at neutral pH. Therefore at neutral pH, the CB[6] ring resides on the tetramethylenediammonium recognition unit so that both portals of the macrocycle are able to engage in ion-dipole binding interactions, thereby blocking the mesopore entrances and encapsulating the guest molecules.

Zink and coworkers have reported the potential application of cyclodextrins (CDs) in the development of pH-sensitive nanovalves (Ambrogio et al. 2011; Coti

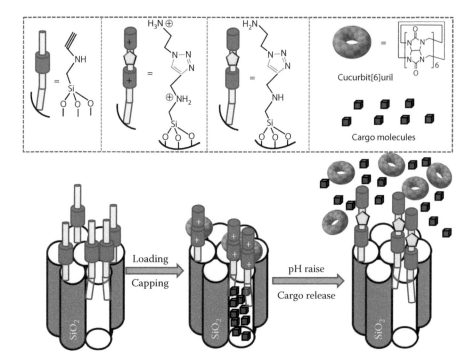

FIGURE 3.3 **(See color insert.)** Graphical representation of the performance of pH-responsive supramolecular valves based in [2]pseudorotaxanes consisting of bisammonium stalks and cucurbit[6]uril (CB[6]) rings. The alkyne-functionalized MCM-41 is loaded with a model molecule and capped with CB[6] during the CB[6]-catalyzed alkyne-azide 1,3-dipolar cyclo-additions, followed by washing away the excess of substrates. Entrapped cargo is released by switching off the ion-dipole interactions between the CB[6] rings and the bisammonium stalks upon raising the pH value. *Source*: Angelos et al. (2008).

et al. 2009; Li et al. 2012; Saha et al. 2007). In 2009, they introduced a new category of MSNPs consisting of hollow mesoporous silica nanoparticles capped by supramolecular machines based on CDs (Du et al. 2009). These authors prepared pH-responsive nanovalves controlled by a supramolecular system containing α-CD rings interacting by hydrogen bonds with anilinoalkane stalks that were tethered to the silica surface (Figure 3.5). When the α-CD rings were complexed with the stalk at neutral pH, the α-CD rings were located near the nanopore openings, blocking departure of cargo molecules that were loaded in the nanopores and hollow interior of the particle. When the nitrogen atoms on the aniline residues were protonated at lower pH, the binding affinities between the α-CD rings and the stalks decreased, releasing the α-CD and allowing the cargo molecules to escape. The same research group also reported a system based on the function of β-CD nanovalves that were also responsive to the endosomal acidification (Meng et al. 2010; Zhao et al. 2010).

In some recent work, a group investigated the opposite recognition; i.e., the β-CD rings were immobilized and the stalks were movable (Meng et al. 2010). In this case, the removable rhodamine B/benzidine stalks acted as nanopistons and

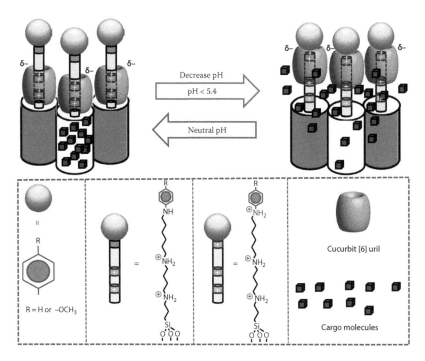

FIGURE 3.4   The pH clock operation of mechanized MSNPs. At neutral pH, the CB[6] ring sits on the tetramethylenediammonium recognition unit, blocking the nanopores. When the pH is lowered below 5.4, the anilinium nitrogen atom is protonated, the CB[6] ring shuttles to the distal hexamethylenediammonium station, and the entrapped cargo is released. The system is designed to allow tuning the pH response by changing R, which modulates the pK_a of the anilinium nitrogen atom. *Source*: Angelos, Khashab et al. (2009).

moved in and out of the cylindrical cavities provided by the β-CD rings in response to changes in pH.

Park et al. have reported the controlled release of molecules from mesoporous silica nanoparticles (MSNPs) by using a pH-sensitive polyethyleneimine/cyclodextrin (PEI/CD) polypseudorotaxane motif (C. Park et al. 2007) (Figure 3.6). The mesopores were filled with a guest molecule (calcein) and then blocked by threading of CD onto the surface-grafted PEI chains at pH 11. At pH 5.5, calcein molecules were released from the pores by the reversible dethreading of CD from the PEI chain.

The incorporation of nanocaps tethered to the mesopore entrances of MSNPs through acid cleavable chemical bonds has been also reported as a good strategy to develop pH-responsive release systems. Thus gold (Aznar et al. 2009) and Fe_3O_4 nanoparticles (Gan et al. 2011) have been used as blocking caps to control the transport of molecules from MSNPs through a reversible pH-dependent boronate ester bond, which is hydrolyzed under acid pH. Acid-labile acetal linker has been also reported as a pH-responsive nanogated ensemble by capping the pores of MSNPs with gold NPs through the formation of these pH-sensitive bonds (R. Liu et al. 2010).

Recently, Chen et al. have developed an ingenious pH-responsive release system based on DNA nanoswitch-controlled organization of gold nanoparticles attached to

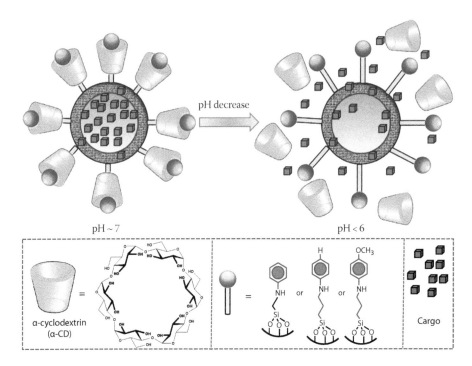

FIGURE 3.5 **(See color insert.)** Schematic representation of pH-responsive operating mechanism of hollow MSNPs capped by supramolecular machines based on α-CD. At pH ~7, the aniline nitrogens on the stalks are not protonated and are encircled by the α-CD rings that block the pores. When the pH is decreased, protonation of the nitrogen causes α-CD to dissociate, unblocking the pores and releasing the trapped cargo molecules. *Source*: Du et al. 2009).

MSNPs (L. Chen, Di et al. 2011). In this system, the hybridization and dehybridization of DNA between strands 1 and 2 were controlled by the adjustment of pH of aqueous media. This structural conversion allows unlocking and closing the mesopore outlets, leading to the controlled release of loaded molecules from pore voids at acid conditions.

Another method consists of covalently linking pH-responsive polymers to the external surface of MSNPs, which act as smart nanoshells able to regulate the release of guest molecules. Thus Liu et al. reported the grafting of poly(4-vinyl pyridine) on the mesoporous silica surface to create a nanoshell that acts as a pH-sensitive barrier controlling the release of the molecules trapped inside the pores (Figure 3.7) (R. Liu et al. 2011).

Poly (acrylic acid) (PAA) grafted MSNPs have been also prepared, demonstrating that the drug release rate was pH dependent and increased with the decrease of pH (Hong, Li, and Pan 2009; Yuan et al. 2011). These pH-responsive nanocontainers exhibit similar release behavior. In alkaline solution, the polymer chains are extended, the mesopore entrances are open, and the drug is released. In contrast, in acidic medium the compact polymer layers block the pores and hinder the drug diffusion out of the channels. However, in certain pathologies such as cancer, where

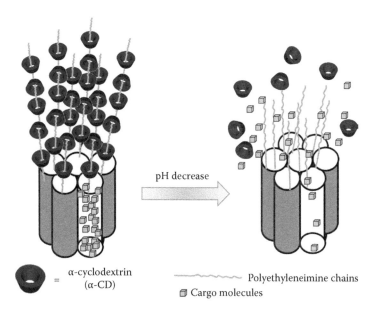

FIGURE 3.6 Schematic representation of the pH-responsive release of molecules from MSNPs modified with PEI/CD polypseudorotaxane. *Source*: Park et al. (2007)

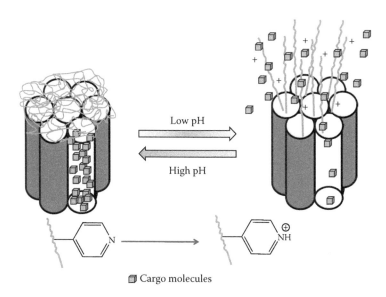

FIGURE 3.7 Schematic representation of the controlled release of molecules from MSNPs coated by the pH-responsive polymer poly(4-vinyl pyridine). The grafted polymer nanoshell can work as a pH-sensitive barrier to control the release of trapped molecules from mesoporous silica. *Source*: Liu et al. (2011).

targeted smart drug delivery is especially relevant, the external pH of cancerous tissues is usually lower than that of the surrounding healthy tissues (Engin et al. 1995). Consequently, suitable cancer therapies demand the development of acid-triggered delivery systems that release the drugs under acidic conditions and impede drug release at physiological pH. Thus Sun et al. have recently developed new pH-responsive nanodevices consisting of MSNPs poly(2-(diethylamino)ethyl methacrylate) (PDEAEMA) grafted to MSNPs (Sun, Hong, and Pan 2010). The tertiary amine in PDEAEMA makes it easy to get a proton to form quaternary ammonium, and the polymer adopts the coil (soluble) conformation in acidic solution, allowing the release of the cargo. Conversely, in neutral or alkaline solution, the polymer is in the collapsed (insoluble) state due to the hydrophobic interaction of polymer chains, impeding the diffusion of the drug out of the pores.

3.2   TEMPERATURE AS RELEASE TRIGGER

In recent years the attachment of thermo-responsive polymers, such as poly(N-isopropylacrylamide) (PNIPAM) and its derivatives, onto the surface of MSNPs has been reported (J.-H. Park, Lee, and Oh 2007; You et al. 2008; S. Zhu et al. 2007; Y. Zhu et al. 2009). PNIPAM transforms its molecular chain conformation in response to temperature in aqueous environments (Fu et al. 2006). Thus the molecular chains of PNIPAM are hydrated below the lower critical solution temperature (LCST) of 32°C, resulting in an extended chain conformation that prevents the departure of the drugs loaded inside the mesopore channels. Increasing the temperature above the LCST dehydrates the polymer chains, resulting in a collapsed conformation, pore opening, and subsequent release of the entrapped cargo.

An increase in the LCST under physiological conditions would be desirable for biomedical applications. This can be achieved by performing modifications of the polymer composition by copolymerization with other monomers such as acrylamide (Nagase et al. 2007; Zintchenko, Ogris, and Wagner 2006) or N-isopropylmethacrylamide (Hoare et al. 2009; Keerl et al. 2008). Very recently, Baeza et al. have presented a novel nanodevice able to perform remotely controlled release of small molecules and proteins in response to an alternating magnetic field (Figure 3.8) (Baeza et al. 2012). This device is based on MSNPs with iron oxide nanocrystals encapsulated inside the silica matrix and decorated on the outer surface with a thermo-responsive copolymer of poly(ethylenimine)-β-poly(N-isopropylacrylamide) (PEI/NIPAM). The polymer structure has been designed with a double purpose, to act as temperature-responsive gatekeeper for the drugs hosted inside the mesopores and to retain proteins in the polymer shell by electrostatic or hydrogen bond interactions. The nanosystem retains the different cargos at low temperature (20°C) and delivers the entrapped molecules when the temperature exceeds 35–40°C following different release kinetics.

Bein et al. described a molecular valve that releases entrapped fluorescein upon heating to the specific melting temperature of double-stranded DNA sequences that are attached to the pore openings of mesoporous nanoparticles (Schlossbauer et al. 2010).

A different approach to synthesizing temperature-responsive drug delivery systems has been recently developed by Martinez-Máñez and coworkers (Aznar et al.

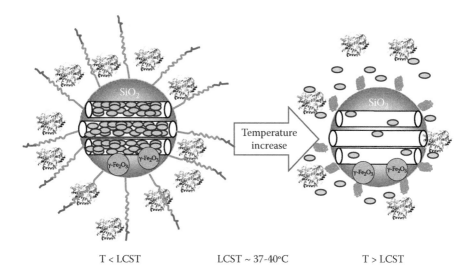

T < LCST   LCST ~ 37-40ºC   T > LCST

FIGURE 3.8   (See color insert.) Schematic depiction of the temperature-responsive performance of MSNPs encapsulating iron oxide nanocrystals and decorated on the surface with a thermo-responsive copolymer of poly(ethyleneimine)-b-poly(N-isopropylacrylamide) (PEI/NIPAM). The polymer acts both as temperature-responsive gatekeeper for the drugs trapped inside the silica matrix and retains proteins in the polymer shell by electrostatic or hydrogen bond interactions. The nanocarrier traps the different cargos at low temperatures (20°C) and releases the retained molecules when the temperature exceeds 35−40°C following different kinetics. *Source*: Baeza et al. (2012).

2011). This strategy consisted of loading MSNPs with a fluorescent cargo and functionalizing with octadecyltrimethoxysilane. The alkyl chains interact with paraffin, which builds a hydrophobic layer around the particle. Upon melting of the paraffin, the guest molecule is released. The release temperature can be tuned by choosing appropriate paraffin.

3.3   REDOX POTENTIAL AS RELEASE TRIGGER

Another interesting approach is to take advantage of the different redox potentials between the intracellular and extracellular space as internal stimuli to trigger the release from MSNPs. Accordingly, different redox potential-responsive release systems have been developed, consisting of using different nanocaps, such as CdS (Lai et al. 2003), Fe_3O_4 (Giri et al. 2005), or Au (Torney et al. 2007) nanoparticles, covalently grafted to the MSNPs through chemically labile disulfide linkages. The removal of the nanocaps and the subsequent release of the entrapped cargo were achieved by cleaving such linkages using disulfide-reducing agents, such as dithiothreitol (DTT) or mercaptoethanol (ME) (Figure 3.9). Liu et al. (R. Liu et al. 2008) reported the use of cross-linked poly(N-acryloxysuccinimide) attached at the pore entrance of MSNPs. After loading the dye molecules into MSNPs, the openings were blocked by the addition of cystamine, a disulfide-based bifunctional primary amine, which allows polymer chains to be cross-linked through the reaction with

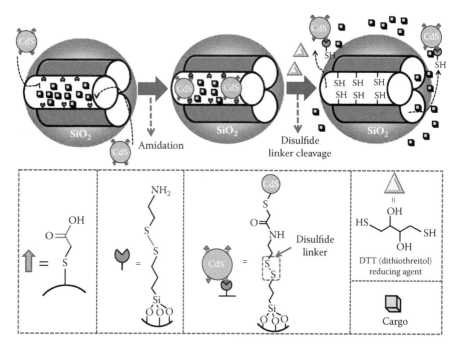

FIGURE 3.9 Schematic depiction of the CdS nanoparticle-capped MSNPs' drug delivery system. The controlled delivery mechanism of the system relies on the chemical reduction of the disulfide linkage between the CdS caps and the MSNPs' host by using a reducing agent. *Source*: Lai et al. (2003).

N-oxysuccimide groups along the polymer chain. The presence of DTT cleaved the disulfide bond of the cystamine, causing a disruption in the polymeric network and leading to the redox potential-driven delivery. More recently, the immobilization of collagen on the outer surface of MSNPs by disulfide bonds, which can be cleaved with various reducing agents, has been also accomplished (Luo et al. 2011).

Zink et al. reported the fabrication of snap-top systems using MSNPs functionalized with rotaxanes incorporating disulfide bonds in their stalks, which are encircled by curcurbit[6]uril or α-CD rings (Ambrogio et al. 2010). Upon exposition to DTT, the reductive cleavage of disulfide bonds in the stalks resulted in the snapping of the stalks of the rotaxanes and led the cargo to release from MSNPs. A similar approach was carried out by Kim et al., which consisted of using glutathione-induced intracellular release of molecules from MSNPs with CD gatekeepers covalently linked onto the particle surface via the disulfide unit (H. J. Kim et al. 2010).

However, different research studies have revealed the low reductive or even oxidative medium in endosomes (Austin et al. 2005). Therefore the contact to the cytosol is essential to cleave the disulfide linkages and release the cargo. The use of photoactive compounds to open the endosome is one prospect for tackling this challenge (de Bruin et al. 2008; Febvay et al. 2010). Bein and coworkers demonstrated that there was no release of disulfide-bound dye from MSNPs endocytozed by cells (Sauer et al. 2010). Inefficient endosomal escape is generally a bottleneck for molecular delivery into the cytoplasm. However, after photochemical rupture of the endosomes by means of a photosensitizer, MSNPs successfully released disulfide-bound dye into the cytoplasm, showing that the reducing milieu of the cytoplasm is enough to cleave the disulfide linkages.

3.4 LIGHT AS RELEASE TRIGGER

Since processes involving light activation are rapid and directional, light-responsive MSNPs permit low invasiveness in biological systems. Fujiwara et al. (Mal, Fujiwara, and Tanaka 2003; Mal et al. 2003) reported for the first time that the uptake, storage, and release of organic molecules in MCM-41 could be regulated through the photo-controlled and reversible intermolecular dimerization of coumarin derivatives attached to the pore entrances. Later on, a multifunctional, fully controlled storage and release system was developed by attaching azobenzene groups on the mesopore outlets (Angelos et al. 2007; Lu et al. 2008; Zhu and Fujiwara 2007). The release of guest molecules included in the pores of mesoporous silica was promoted by simultaneous irradiation with UV (ultraviolet) and VIS (visible) light, which made the azobenzene molecules act as both impellers and nanogates (Figure 3.10). The reversible *cis-trans* photoisomerization of azobenzene substituents on the pore surface by a rotation-inversion mechanism caused a stirring action that accelerates the diffusion of the guest from the mesopores.

Zink and coworkers have reported the use of light-operated mechanized MSNPs whose operation is based on the high binding affinity in aqueous solution between β-CD and trans-azobenzene derivatives and low, if any, binding between β-CD

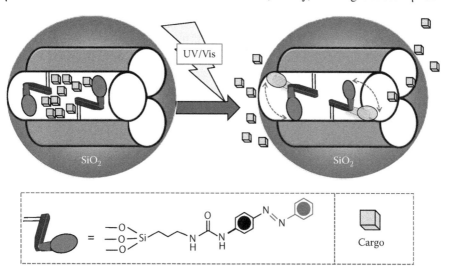

FIGURE 3.10   Nanoimpeller-MSNPs with 4-phenylazoaniline incorporated into the mesopores and loaded with cargo molecules. Irradiation at 450 nm induces *trans*-to-*cis* and *cis*-to-*trans* isomerizations of the azobenzenes, generating a nanoimpeller-type motion that causes the cargo molecules to be propelled out of the mesopores. *Source*: Angelos et al. 2007).

and *cis*-azobenzene derivatives (Ferris et al. 2009). Irradiating with 351 nm light causes the isomerization of azobenzene to the *cis* conformation and therefore pore uncapping. An alternative strategy using CDs was proposed by Park et al., which consisted of covalently linking β-CD on the surface of MSNP through a photo-cleavable o-nitrobenzyl ester moiety (Park, Lee, and Kim 2009). Upon exposure to UV light, the guest molecules were released from the pore by removal of the CD gatekeeper.

Lin and coworkers described the use of gold nanoparticles as light-sensitive gate-keepers by the surface attachment through the photo-responsive linker thioundecyl-tetraethyleneglycoestero-nitrobenzylethyldimethylammonium bromide (TUNA) (Vivero-Escoto et al. 2009). Upon UV irradiation, TUNA would lead to the nega-tively charged thioundecyltetraethyleneglycolcarboxylate (TUEC), leading to the dis-sociation of the Au NPs from the MSNP surface due to charge repulsion. Thus there was an uncapping of the mesopores and the release of guest molecules took place.

In a recent study, sulforhodamine 101 was loaded inside the mesopores of mercaptopropyl-functionalized MSNPs, and the cargo molecules were entrapped by the presence of $Ru(bpy)_2(PPh_3)$-moieties, coordinated to mercaptopropyl func-tional groups (Knezevic, Trewyn, and Lin 2011). Upon irradiation with visible light ($\lambda = 455$ nm), Ru-S coordination bond was cleaved, triggering the release of capping species and loaded molecules.

Another innovative approach consists of anchoring a light-responsive polymer on the pore outlets of MSNPs (Lai et al. 2010). This strategy relies on the fact that the incorporation of hydrophobic or hydrophilic monomers into the PNIPAM backbone can lead to a decrease or increase in the LCST, respectively (Figure 3.11). Therefore when monomers bearing a light-responsive moiety such as azobenzene and 2-nitro-benzyl groups were incorporated in the PNIPAM backbone, the resultant polymers were light-responsive. Their LCST could be easily modulated by applying UV irra-diation since the polarity of the light-responsive moieties changes under irradiation. Thus UV irradiation provokes a change in the polymer conformation from collapsed to coil state, and therefore the gate is opened and subsequently the entrapped mol-ecules escape out ofthe pores.

3.5 MAGNETIC FIELD AS RELEASE TRIGGER

Chen et al. have reported the capping of MSNPs with Fe_3O_4 magnetic nanoparticles (P. Chen, Hu et al. 2011). For this purpose, MSNPs were functionalized with 3-ami-nopropyltrimethoxysilane (APTS) and then loaded with the antitumor drug camp-tothecin (CPT). The mesopore entrances with the CPT-loaded amine-MSNPs were covalently capped through amidation of the APTES bound at the pore surface with *meso*-2,3-dimercaptosuccinic acid functionalized superparamagnetic iron oxide magnetic nanoparticles with an average diameter of 5.6 nm. Under the application of a magnetic trigger, the Fe_3O_4 nanocaps were removed due to the cleavage of chemical bonds and that subsequently led to a fast-response drug release.

Very recently, Vallet-Regí and coworkers (Ruiz-Hernández, Baeza, and Vallet-Regí 2011) have used alternating magnetic fields as an external release trigger to

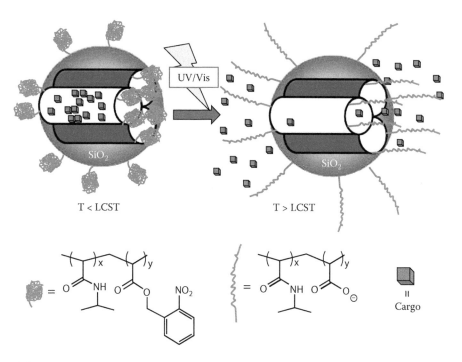

FIGURE 3.11   Light-responsive delivery system based on responsive polymer grafted to MSNPs. *Source*: Lai et al. (2010).

develop "on-off" stimuli-responsive drug delivery systems. The system is based on MSNPs encapsulating iron oxide superparamagnetic nanocrystals (γ-Fe_2O_3), which provide the potential to perform targeting and magnetic resonance imaging. The MSNP's surface was functionalized with a single-stranded DNA, selected to display a melting temperature with its complementary strand of 47°C, which is in agreement with the upper limit of hyperthermia treatment. Then mesoporous channels were loaded with fluorescein, chosen as the model molecule (Figure 3.12).

To design suitable nanocaps, a different batch of iron oxide γ-Fe_2O_3 superparamagnetic nanocrystals with average size similar to the mesopore diameter (ca. 5 nm) were synthesized and functionalized with the complementary DNA strand, acting then as gatekeepers upon hybridization between both strands (Figure 3.13) (Ruiz-Hernández, Baeza, and Vallet-Regí 2011).

An appropriate stimulus then should be chosen to trigger the progressive double-stranded DNA dehybridization, which would give rise to pore uncapping and subsequent release of the entrapped cargo (Figure 3.14) (Ruiz-Hernández, Baeza, and Vallet-Regí 2011). The magnetic-responsive release of this system relies on the thermo-sensitive oligonucleotide linkage. Therefore the stimuli-responsive capability of this system was tested by exposition of the drug-loaded MSNPs to an alternating magnetic field, which led to a temperature increase. Once the temperature reached 47°C, the pore channels were uncapped and the cargo molecule was released.

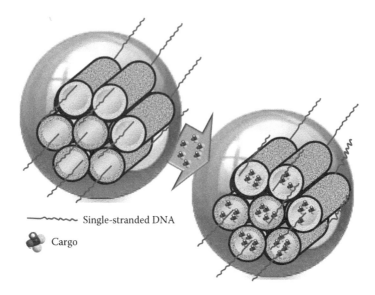

FIGURE 3.12   Magnetic MSNPs modified with a specific single-stranded DNA and loaded with a drug molecule. *Source*: Ruiz-Hernández, Baeza, and Vallet-Regí (2011).

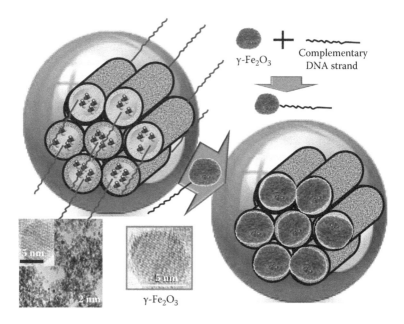

FIGURE 3.13   Schematic representation of magnetic MSNPs modified with the single-stranded DNA and nanogates consisting of γ-Fe₂O₃ nanoparticles modified with the complementary DNA strand. The hybridization of both DNA strands provokes the mesopore capping. *Source*: Ruiz-Hernández, Baeza, and Vallet-Regí (2011).

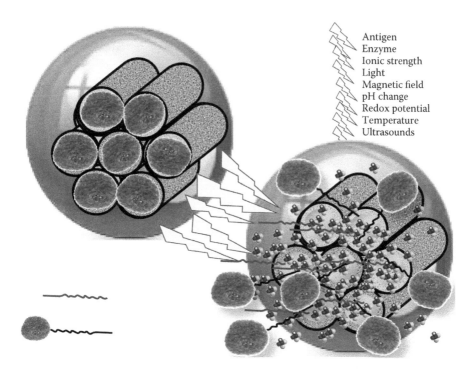

Antigen
Enzyme
Ionic strength
Light
Magnetic field
pH change
Redox potential
Temperature
Ultrasounds

FIGURE 3.14 **(See color insert.)** Progressive double-stranded DNA dehybridization upon the application of an appropriate stimulus. In this case, the removal of the DNA/magnetic nanocap conjugates and the subsequent cargo release take place as a result of temperature increase due to the application of an alternating magnetic field. *Source*: Ruiz-Hernández, Baeza, and Vallet-Regí (2011).

One noticeable feature of this system is the reversibility of the DNA linkage, which results in an "on-off" release mechanism. Thus the purpose of the nanocapping system here is to act as a reversible gatekeeper; i.e., cargo release is triggered when the temperature is raised, whereas molecule release is hindered when the temperature is stabilized (Figure 3.15). This "on-off" behavior makes this system a potential on-demand drug delivery device (Ruiz-Hernández, Baeza, and Vallet-Regí 2011).

Moreover, the magnetic component of the whole system allows hyperthermia temperatures (42–47°C) to be reached under an alternating magnetic field (Figure 3.16) (Ruiz-Hernández, Baeza, and Vallet-Regí 2011). All these features provide this system with attractive potential biomedical applications such as the synergistic combination of drug delivery and hyperthermia treatments for cancer therapy.

3.6   ULTRASOUND AS RELEASE TRIGGER

Ultrasound can be used as a noninvasive external trigger for the pulsatile release of drugs from mesoporous silica. Ultrasound can penetrate deep into the interior of the body, and it can be carefully controlled through a number of parameters including frequency, power density, duty cycles, and time of application (Gao, Fain, and

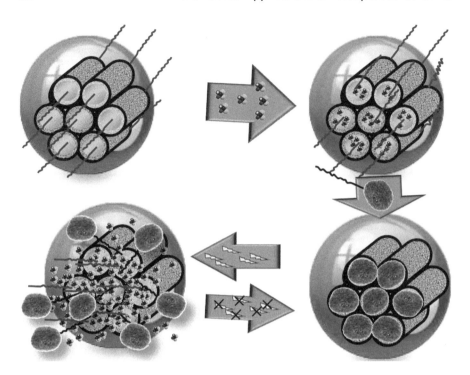

FIGURE 3.15 Reversible magnetic nanodevices through iron oxide/DNA gates. *Source*: Ruiz-Hernández, Baeza, and Vallet-Regí (2011).

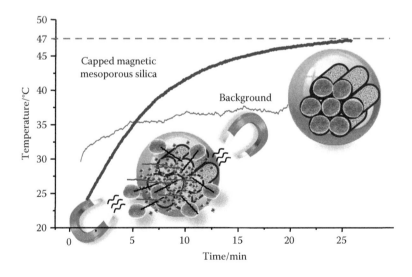

FIGURE 3.16 Hyperthermia assay on capped magnetic mesoporous silica under an alternating magnetic field of 24 kA·3m^{-1} and 100 kHz. *Source*: Ruiz-Hernández, Baeza, and Vallet-Regí (2011).

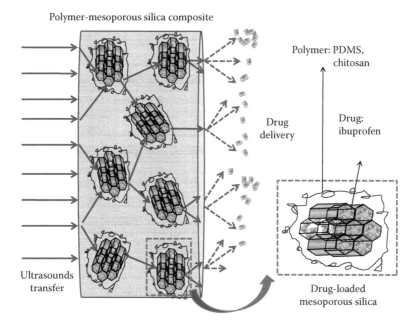

Polymer-mesoporous silica composite

Polymer: PDMS, chitosan

Drug delivery

Drug: ibuprofen

Ultrasounds transfer

Drug-loaded mesoporous silica

FIGURE 3.17   Polymer-mesoporous silica hybrid composites as ultrasound-triggered drug delivery devices. The tested polymers have been poly(dimethylsiloxane) (PDMS) (Kim et al. 2006) and chitosan (Depan, Saikia, and Singh 2010).

Rapoport 2005). Kim et al. (H. J. Kim et al. 2006) reported the development of poly(dimethylsiloxane) (PDMS)-mesoporous silica composite as an ultrasound-triggered smart drug delivery device. More recently, Depan et al. synthesized novel chitosan-mesoporous silica hybrid composites as ultrasound-triggered drug delivery devices (Depan, Saikia, and Singh 2010). In both cases, the ultrasound drug release is due to the cavitation effect without causing any significant destruction on the polymer morphology (Figure 3.17). These findings suggest the feasibility of ultrasounds for the pulsatile release of molecules from mesoporous silica matrices.

3.7   ENZYMES AS RELEASE TRIGGERS

The design of enzyme-responsive nanocaps blocking the pores of MSNPs is another attractive prospect since there is an anomalous increase of enzymatic presence or activity in some diseased tissues. Zink and coworkers functionalized the pore outlets of MSNPs with a [2]rotaxane capped with an ester-linked adamantyl stopper (Patel et al. 2008). The system released its cargo after addition of porcine liver esterase, which induced dethreading of the [2]rotaxane due to hydrolysis of the adamantyl ester. CDs have been also attached on the MSNP surface via "click chemistry" reactions. The addition of α-amilase catalyzed the hydrolysis of these groups, allowing the release of molecules entrapped into the mesopores (C. Park et al. 2009). Bein and coworkers reported the attachment of avidin caps on biotinylated MSNPs (Schlossbauer, Kecht, and Bein 2009). The addition of the protease trypsin provoked the hydrolysis of the attached avidin and the release of the entrapped cargo. Martínez-Máñez and coworkers described the capping of silica mesoporous supports with lactose and the selective uncapping in the presence of enzyme β-D-galactosidase (Figure 3.18) (Bernardos et al. 2009). The induced delivery in the presence of galactosidase is due to the enzymatic facilitated rupture of the glycosidic bond and size reduction of the saccharide chain at the surface. The same research group functionalized the silica mesoporous support's surface with saccharide derivatives, and the cargo release was achieved by enzymatic hydrolysis in the presence of pancreatin or β-D-galactosidase in pure water, and also in intracellular media by lysosomal enzymes (Bernardos et al. 2010).

Very recently, the same research group designed multi-enzyme-responsive capped MSNPs using bulky organic moieties containing amide and urea linkages to block the mesopores (Agostini et al. 2012). MSNPs released their cargo upon addition of amidase and urease. Amidase induced an immediate, yet not complete, release of the cargo. On the other hand, urease allowed a near total cargo release that was delayed in time. Thus the possibility of including in the capping molecule different enzyme-hydrolyzable groups located in predefined positions allows control of the delivery profiles.

Enzyme-responsive polymers have been also linked to MSNPs to provide them with bioresponsive drug delivery capability using the antitumor drug doxorubicin

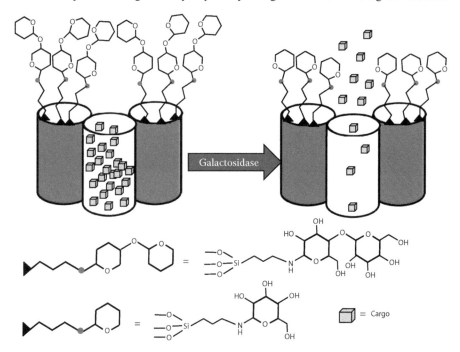

FIGURE 3.18 Scheme of silica mesoporous supports entrapping cargo molecules and capped with trialkoxysilane lactose derivative. The addition of the enzyme β-D-galactosidase triggers the release of the loaded cargo. *Source*: Bernardos et al. (2009).

(Singh et al. 2011). The authors first electrostatically adsorbed an acrylamide to the MSNP surface and then used the acryl groups to synthesize a covalently cross-linked PEG-based polymer shell. The authors characterized MSNPs' doxorubicin-eluting properties in vitro and demonstrate that the polymer-coated MSNPs release the entrapped drug in response to proteases present at a tumor site in vivo, resulting in cellular apoptosis.

3.8 ANTIGENS AS RELEASE TRIGGERS

A revolutionary strategy consists of using the highly specific antibody-antigen interaction as a powerful switchable method to develop stimuli-responsive silica mesoporous materials for controlled release functions.

Thus Martínez-Máñez and coworkers reported the functionalization of the pore outlets of MSNPs with a certain hapten able to be recognized by an antibody that acts as a nanoscopic cap (Figure 3.19) (Climent et al. 2009). The opening protocol and delivery of the entrapped cargo rely on the highly effective displacement reaction involving the presence in the solution of the antigen to which the antibody is selective.

3.9 APTAMER TARGET AS RELEASE TRIGGER

Nucleic acid aptamers are single-stranded short oligonucleotide sequences that can bind specific targets with high affinity and specificity (Ellington and Szostak 1990; Mairal et al. 2008; Shamah, Healy, and Cload 2008; Tuerk and Gold 1990). In comparison with antibodies, aptamers, especially DNA aptamers, are relatively easier to obtain, more stable to biodegradation, less susceptible to denaturation, and flexible to modification. All these aspects make aptamers ideal candidates for designing novel aptamer-target-responsive mesoporous silica for smart drug delivery applications (Özalp et al. 2011).

The earliest research activities regarding gated mesoporous materials using nucleotides were carried out by Climent et al., who used a single-stranded DNA oligonucleotide to cap the mesopore entrances (Climent et al. 2010). For this purpose, they functionalized the external surface of cargo-loaded MSNPs with aminopropyl groups. These amino groups create a positively charged compound at neutral pH in water and interact with negatively charged oligonucleotides, resulting in the closing of the mesopores. The loaded cargo was released in the presence of the target complementary strand due to the hybridization of the two oligonucleotides, which triggers the uncapping of the pores (Figure 3.20) (Climent et al. 2010).

Although complementary DNA-triggered delivery does not present many real applications, this pioneer report proves that nucleic acids can be used to design biogated delivery systems that can be selectively opened in the presence of specific targets.

Later on, Zhu et al. designed a novel stimulus-responsive silica mesoporous system based in aptamer-target interactions (C.-L. Zhu et al. 2011). This system is based on MSNPs whose pores were capped with Au nanoparticles modified with aptamer (adenosine triphosphate [ATP] aptamer in this case). In the presence of ATP molecules, the Au nanoparticles were uncapped by a competitive displacement reaction

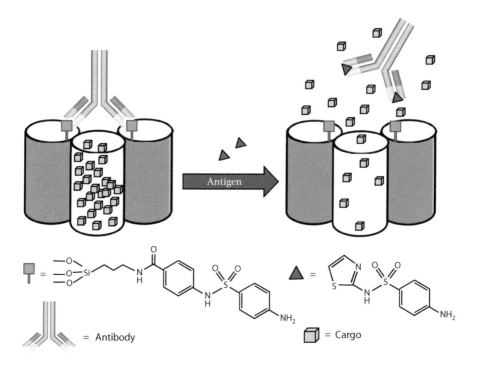

FIGURE 3.19   Scheme of silica mesoporous scaffold modified with hapten (4-(4-aminoben-zenesulfonylamino)benzoic acid) derivative and capped with the polyclonal antibody for sulfathiazole. The addition of the antigen sulfathiazole triggers cargo release. *Source*: Climent et al. (2009).

and the entrapped cargo was released. The operation mechanism of the system is illustrated in Figure 3.21. MSNPs were functionalized with aminopropyl groups, and a derivative of the ATP molecule (adenosine-50-carboxylic acid, denoted as adenosine-COOH) was immobilized on the external surface of mesoporous silica through amidation reaction. Meanwhile, Au nanoparticles were functionalized with ATP aptamer through Au-S bond. Since the ATP aptamer recognizes the adenine and ribose moieties and not the phosphate moiety (Dieckmann et al. 1997) when mixing aptamer-modified Au nanoparticles with adenosine-modified MSNPs, Au nanoparticles would cap the mesopores owing to the binding reaction of ATP aptamer with adenosine. The addition of ATP resulted in a competitive displacement reaction to the adenosine-aptamer interaction, leading to the mesopores' uncapping and delivery of cargo molecule.

Recently, Özalp et al. developed switchable controlled drug delivery systems using aptamers as nanovalves (Özalp and Schäfer 2011). For the proof of principle, MSNPs were modified with a hairpin form of ATP-binding DNA, which allowed encapsulation and release of guest molecules controlled by ATP. The operation mechanism of the drug delivery system centered on the conversion of an aptamer sequence into a molecular beacon-type hairpin structure, which was

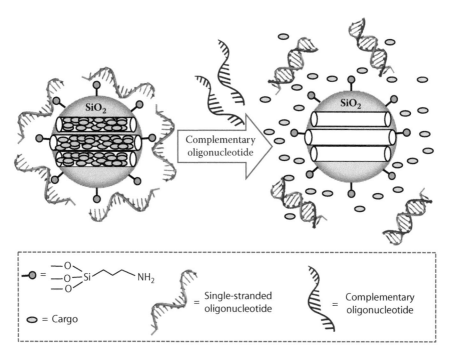

FIGURE 3.20 Single-stranded DNA oligonucleotide capping of MSNPs and stimulus-responsive cargo release upon addition of the complementary oligonucleotide strand. *Source*: Climent et al. (2010).

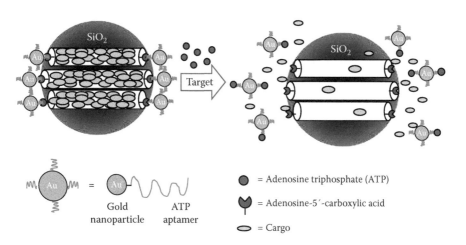

FIGURE 3.21 MSNPs functionalized with adenosine-5′-carboxylic acid, loaded with cargo and end-capped with Au nanoparticles modified with the ATP aptamer. Pore opening and subsequent cargo release are triggered by effective displacement reaction in the presence of the target molecule (ATP). *Source*: Zhu et al. (2011).

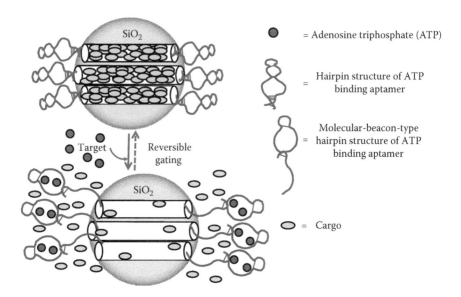

FIGURE 3.22 Schematic representation of the reversible aptamer-target-responsive controlled drug delivery using aptamer-based switchable nanovalves. *Source*: Özalp and Schäfer (2011).

used as a nanovalve. The hairpin of the aptamer sequences had originally been designed for signaling the presence of target molecules. The signaling is usually based on a duplex-to-complex intramolecular switching, resulting in denaturation of the duplex region (Vasir, Reddy, and Labhasetwar 2005). Most of the aptamer sequences can be converted into such a structure-switching hairpin design, where the hairpin switches between a single-stranded DNA structure and a duplex DNA in the region of the molecule that is close to the surface. Since a duplex DNA helix has the dimensions of ca. 2 nm thickness and single-stranded DNA is of subnanometer size, DNA duplexes can be used to cap mesopores of 2.6 nm in diameter to avoid premature cargo release. The duplex neck region of the hairpin structure of the ATP-binding aptameras used in this study has similar dimensions to those of the aforementioned examples of duplex DNA, which makes them suitable to cap the pores of MSNPs (with average pore sizes in the 2.3–2.7 nm range) (Figure 3.22) (Özalp and Schäfer 2011). The developed system was functional, as the hairpin aptamer blocked the pores prior to addition of ATP, whereas the presence of ATP specifically triggered their opening and consequently the cargo (fluorescein) release. In addition, the reversibility of the aptamer-gate-keeping function, and hence the on-off capability of the nanovalves, was evaluated and confirmed by repeatedly adding and removing ATP while monitoring the concentration of fluorescein released. The reversibility of the system provides it with improved properties, and the partial delivery of drug molecules can be controlled better instead of a sudden release as is the case in snap-top designs.

3.10 DUAL STIMULI AS RELEASE TRIGGERS

Dual-controlled or multiresponsive controlled release systems are those able to respond to two or more inputs either in an independent or in a synergistic way. The research group of Martínez-Máñez reported the attachment of suitable polyamines on the MSNP surface to obtain dual stimuli-responsive gate-like ensembles. The release of the cargo entrapped inside the mesoporous matrix was achieved by employing pH modifications or in the presence of certain anions (Casasús et al. 2008). The same research group also described the functionalization of the pore outlets of MSNPs with boronic acid functionalized gold nanoparticles acting as nanocaps. These Au nanoparticles were linked to the surface of saccharide-functionalized MSNPs through the formation of boronate ester bonds that are hydrolyzed under acid conditions (pH = 3). The nanosystem exhibited an "on-off" response due to the reversibility of the boronate bond formation. Furthermore, the metallic nanoparticles could be heated by laser irradiation at 1064 nm (266 mJ), causing a plasmon resonance light-induction release due to the thermal boronate bond cleavage. Angelos et al. demonstrated that dual-stimuli controlled-release MSNPs could also be used as AND logic gates (Angelos, Yang et al. 2009). These molecular machines were designed to operate in tandem with one another in such a way that the dual-controlled nanoparticle systems function as AND logic gates. In this case, two different molecular machines were mounted on the mesoporous silica surface, azobenzene as light-activated nanoimpellers and [2]pseudorotaxanes as pH-responsive nanovalves. The two systems can act separately, but only the simultaneous activation of both molecular machines conducts to the load release, showing that these kind of devices not only can be used in drug delivery applications but can perform simple logic operations. Liu et al. reported a multiresponsive supramolecular capped mesoporous silica system by grafting β-CD-bearing polymer on the surface of mesoporous silica and cross-linking by the addition of disulfide groups to form a polymeric network that blocked the pore entrances (J. Liu et al. 2009). The nanodevice was able to release its cargo in response to three different stimuli: UV light (causing the isomeric transformation of azobenzene groups), presence of α-CD (as competitive ligands to displace β-CD), and the addition of disulfide reductive agents such as DTT (to cleave the disulfide bond between β-CD and polymer main chains). In a very recent work, Chen et al. described the synthesis of dual stimuli-responsive vehicles by attaching self-complementary duplex DNA to the pore openings of MSNPs, which resulted in a cap for trapping guest molecules (Chen, Geng et al. 2011). The duplex DNA cap could be either denatured by heating or hydrolyzed by endonucleases, thus opening the nanopores and releasing the cargo.

Very recently, Chang et al. reported the grafting of dual-stimuli-responsive polymers to MSNPs (Chang et al. 2011). The authors synthesized core-shell MSNPs with thermo/pH-coupling sensitivity. MSNPs were used as the core and cross-linked poly(N-isopropylacrylamide-co-methacrylic acid) (P(NIPAM-co-MAA)) polymer was used for the outer shell. The thermo-sensitive volume phase transition (VPTT) could be precisely regulated by the molar ratios of MAA to NIPAM and the concentration of NaCl. The amount of drug released was small below the VPTT and increased above such value, exhibiting an apparent thermo/pH-response controlled drug release.

REFERENCES

Agostini, Alessandro, Laura Mondragón, Carmen Coll, Elena Aznar, M. Dolores Marcos, Ramón Martínez-Máñez, Félix Sancenón, Juan Soto, Enrique Pérez-Payá, and Pedro Amorós. 2012. "Dual Enzyme-Triggered Controlled Release on Capped Nanometric Silica Mesoporous Supports." *ChemistryOpen* 1 (1):17–20.

Ambrogio, Michael W., Travis A. Pecorelli, Kaushik Patel, Niveen M. Khashab, Ali Trabolsi, Hussam A. Khatib, Youssry Y. Botros, Jeffrey I. Zink, and J. Fraser Stoddart. 2010. "Snap-Top Nanocarriers." *Organic Letters* 12 (15):3304–7.

Ambrogio, Michael W., Courtney R. Thomas, Yan-Li Zhao, Jeffrey I. Zink, and J. Fraser Stoddart. 2011. "Mechanized Silica Nanoparticles: A New Frontier in Theranostic Nanomedicine." *Accounts of Chemical Research* 44 (10):903–13.

Angelos, Sarah, Eunshil Choi, Fritz Vogtle, Luisa De Cola, and Jeffrey I. Zink. 2007. "Photo-Driven Expulsion of Molecules from Mesostructured Silica Nanoparticles." *Journal of Physical Chemistry C Nanomater Interfaces* 111 (18):6589–92.

Angelos, Sarah, Niveen M. Khashab, Ying-Wei Yang, Ali Trabolsi, Hussam A. Khatib, J. Fraser Stoddart, and Jeffrey I. Zink. 2009. "pH Clock-Operated Mechanized Nanoparticles." *Journal of the American Chemical Society* 131 (36):12912–14.

Angelos, Sarah, Ying-Wei Yang, Niveen M. Khashab, J. Fraser Stoddart, and Jeffrey I. Zink. 2009. "Dual-Controlled Nanoparticles Exhibiting AND Logic." *Journal of the American Chemical Society* 131 (32):11344–46.

Angelos, Sarah, Ying-Wei Yang, Kaushik Patel, J. Fraser Stoddart, and Jeffrey I. Zink. 2008. "pH-Responsive Supramolecular Nanovalves Based on Cucurbit[6]uril Pseudorotaxanes." *Angewandte Chemie International Edition* 47 (12):2222–26.

Ashley, Carlee E., Eric C. Carnes, Genevieve K. Phillips, David Padilla, Paul N. Durfee, Page A. Brown, Tracey N. Hanna, Juewen Liu, Brandy Phillips, Mark B. Carter, Nick J. Carroll, Xingmao Jiang, Darren R. Dunphy, Cheryl L. Willman, Dimiter N. Petsev, Deborah G. Evans, Atul N. Parikh, Bryce Chackerian, Walker Wharton, David S. Peabody, and C. Jeffrey Brinker. 2011. "The Targeted Delivery of Multicomponent Cargos to Cancer Cells by Nanoporous Particle-Supported Lipid Bilayers." *Nat Mater* 10 (5):389–97.

Austin, Cary D., Xiaohui Wen, Lewis Gazzard, Christopher Nelson, Richard H. Scheller, and Suzie J. Scales. 2005. "Oxidizing Potential of Endosomes and Lysosomes Limits Intracellular Cleavage of Disulfide-Based Antibody-Drug Conjugates." *Proceedings of the National Academy of Sciences of the United States of America* 102 (50):17987–92.

Aznar, Elena, Carmen Coll, M. Dolores Marcos, Ramón Martínez-Máñez, Félix Sancenón, Juan Soto, Pedro Amorós, Joan Cano, and Eliseo Ruiz. 2009. "Borate-Driven Gatelike Scaffolding Using Mesoporous Materials Functionalised with Saccharides." *Chemistry: A European Journal* 15 (28):6877–88.

Aznar, Elena, Laura Mondragón, José V. Ros-Lis, Félix Sancenón, M. Dolores Marcos, Ramón Martínez-Máñez, Juan Soto, Enrique Pérez-Payá, and Pedro Amorós. 2011. "Finely Tuned Temperature-Controlled Cargo Release Using Paraffin-Capped Mesoporous Silica Nanoparticles." *Angewandte Chemie International Edition* 50 (47):11172–75.

Baeza, Alejandro, Eduardo Guisasola, Eduardo Ruiz-Hernández, and María Vallet-Regí. 2012. "Magnetically Triggered Multidrug Release by Hybrid Mesoporous Silica Nanoparticles." *Chemistry of Materials* 24 (3):517–24.

Bernardos, Andrea, Elena Aznar, María Dolores Marcos, Ramón Martínez-Máñez, Félix Sancenón, Juan Soto, José Manuel Barat, and Pedro Amorós. 2009. "Enzyme-Responsive Controlled Release Using Mesoporous Silica Supports Capped with Lactose." *Angewandte Chemie International Edition.*

Bernardos, Andrea, Laura Mondragón, Elena Aznar, M. Dolores Marcos, Ramón Martínez-Máñez, Félix Sancenón, Juan Soto, José Manuel Barat, Enrique Pérez-Payá, Carmen Guillem, and Pedro Amorós. 2010. "Enzyme-Responsive Intracellular Controlled Release Using Nanometric Silica Mesoporous Supports Capped with 'Saccharides'." *ACS Nano* 4 (11):6353–68.

Casasús, Rosa, Estela Climent, Ma Dolores Marcos, Ramón Martínez-Máñez, Félix Sancenón, Juan Soto, Pedro Amorós, Joan Cano, and Eliseo Ruiz. 2008. "Dual Aperture Control on pH- and Anion-Driven Supramolecular Nanoscopic Hybrid Gate-Like Ensembles." *Journal of the American Chemical Society* 130 (6):1903–17.

Chang, Baisong, Xianyi Sha, Jia Guo, Yunfeng Jiao, Changchun Wang, and Wuli Yang. 2011. "Thermo and pH Dual Responsive, Polymer Shell Coated, Magnetic Mesoporous Silica Nanoparticles for Controlled Drug Release." *Journal of Materials Chemistry* 21 (25):9239–47.

Chen, Cuie, Jie Geng, Fang Pu, Xinjian Yang, Jinsong Ren, and Xiaogang Qu. 2011. "Polyvalent Nucleic acid/Mesoporous Silica Nanoparticle Conjugates: Dual Stimuli-Responsive Vehicles for Intracellular Drug Delivery." *Angewandte Chemie International Edition* 50 (4):882–86.

Chen, Linfeng, Jiancheng Di, Changyan Cao, Yong Zhao, Ying Ma, Jia Luo, Yongqiang Wen, Weiguo Song, Yanlin Song, and Lei Jiang. 2011. "A pH-Driven DNA Nanoswitch for Responsive Controlled Release." *Chemical Communications* 47 (10):2850–52.

Chen, Po-Jung, Shang-Hsiu Hu, Chi-Sheng Hsiao, You-Yin Chen, Dean-Mo Liu, and San-Yuan Chen. 2011. "Multifunctional Magnetically Removable Nanogated Lids of Fe3O4-Capped Mesoporous Silica Nanoparticles for Intracellular Controlled Release and MR Imaging." *Journal of Materials Chemistry* 21 (8):2535–43.

Climent, Estela, Andrea Bernardos, Ramón Martínez-Máñez, Angel Maquieira, Maria Dolores Marcos, Nuria Pastor-Navarro, Rosa Puchades, Félix Sancenón, Juan Soto, and Pedro Amorós. 2009. "Controlled Delivery Systems Using Antibody-Capped Mesoporous Nanocontainers." *Journal of the American Chemical Society* 131 (39):14075–80.

Climent, Estela, Ramón Martínez-Máñez, Félix Sancenón, María D. Marcos, Juan Soto, Angel Maquieira, and Pedro Amorós. 2010. "Controlled Delivery Using Oligonucleotide-Capped Mesoporous Silica Nanoparticles." *Angewandte Chemie International Edition* 49 (40):7281–83.

Coti, Karla K., Matthew E. Belowich, Monty Liong, Michael W. Ambrogio, Yuen A. Lau, Hussam A. Khatib, Jeffrey I. Zink, Niveen M. Khashab, and J. Fraser Stoddart. 2009. "Mechanised Nanoparticles for Drug Delivery." *Nanoscale* 1 (1):16–39.

Day, Anthony, Alan P. Arnold, Rodney J. Blanch, and Barry Snushall. 2001. "Controlling Factors in the Synthesis of Cucurbituril and Its Homologues." *The Journal of Organic Chemistry* 66 (24):8094–8100.

de Bruin, K. G., C. Fella, M. Ogris, E. Wagner, N. Ruthardt, and C. Bräuchle. 2008. "Dynamics of Photoinduced Endosomal Release of Polyplexes." *Journal of Controlled Release* 130 (2):175–82.

Depan, Dilip, Lakshi Saikia, and Raj Pal Singh. 2010. "Ultrasound-Triggered Release of Ibuprofen from a Chitosan-Mesoporous Silica Composite: A Novel Approach for Controlled Drug Release." *Macromolecular Symposia* 287 (1):80–88.

Dieckmann, Thorsten, Samuel E. Butcher, Mandana Sassanfar, Jack W. Szostak, and Juli Feigon. 1997. "Mutant ATP-Binding Rna Aptamers Reveal the Structural Basis for Ligand Binding." *Journal of Molecular Biology* 273 (2):467–78.

Du, Li, Shijun Liao, Hussam A. Khatib, J. Fraser Stoddart, and Jeffrey I. Zink. 2009. "Controlled-Access Hollow Mechanized Silica Nanocontainers." *Journal of the American Chemical Society* 131 (42):15136–42.

Ellington, Andrew D., and Jack W. Szostak. 1990. "In Vitro Selection of RNA Molecules That Bind Specific Ligands." *Nature* 346 (6287):818–22.

Engin, K., D. B. Leeper, J. R. Cater, A. J. Thistlethwaite, L. Tupchong, and J. D. McFarlane. 1995. "Extracellular pH Distribution in Human Tumours." *International Journal of Hyperthermia* 11 (2):211–16.

Febvay, Sébastien, Davide M. Marini, Angela M. Belcher, and David E. Clapham. 2010. "Targeted Cytosolic Delivery of Cell-Impermeable Compounds by Nanoparticle-Mediated, Light-Triggered Endosome Disruption." *Nano Letters* 10 (6):2211–19.

Ferris, Daniel P., Yan-Li Zhao, Niveen M. Khashab, Hussam A. Khatib, J. Fraser Stoddart, and Jeffrey I. Zink. 2009. "Light-Operated Mechanized Nanoparticles." *Journal of the American Chemical Society* 131 (5):1686–88.

Fu, Qiang, G. V. Rama Rao, Timothy L. Ward, Yunfeng Lu, and Gabriel P. Lopez. 2006. "Thermoresponsive Transport through Ordered Mesoporous Silica/PNIPAM Copolymer Membranes and Microspheres." *Langmuir* 23 (1):170–74.

Gan, Qi, Xunyu Lu, Yuan Yuan, Jiangchao Qian, Huanjun Zhou, Xun Lu, Jianlin Shi, and Changsheng Liu. 2011. "A Magnetic, Reversible pH-Responsive Nanogated Ensemble Based on Fe3O4 Nanoparticles-Capped Mesoporous Silica." *Biomaterials* 32 (7):1932–42.

Gao, Zhong-Gao, Heidi D. Fain, and Natalya Rapoport. 2005. "Controlled and Targeted Tumor Chemotherapy by Micellar-Encapsulated Drug and Ultrasound." *Journal of Controlled Release* 102 (1):203–22.

Giri, Supratim, Brian G. Trewyn, Michael P. Stellmaker, and Victor S. Y. Lin. 2005. "Stimuli-Responsive Controlled-Release Delivery System Based on Mesoporous Silica Nanorods Capped with Magnetic Nanoparticles." *Angewandte Chemie International Edition* 44 (32):5038–44.

Hoare, Todd, Jesus Santamaria, Gerardo F. Goya, Silvia Irusta, Debora Lin, Samantha Lau, Robert Padera, Robert Langer, and Daniel S. Kohane. 2009. "A Magnetically Triggered Composite Membrane for On-Demand Drug Delivery." *Nano Letters* 9 (10):3651–57.

Hong, Chun-Yan, Xin Li, and Cai-Yuan Pan. 2009. "Fabrication of Smart Nanocontainers with a Mesoporous Core and a pH-Responsive Shell for Controlled Uptake and Release." *Journal of Materials Chemistry* 19 (29):5155–60.

Keerl, Martina, Vytautas Smirnovas, Roland Winter, and Walter Richtering. 2008. "Copolymer Microgels from Mono- and Disubstituted Acrylamides: Phase Behavior and Hydrogen Bonds." *Macromolecules* 41 (18):6830–36.

Kim, Hyun-Jong, Hirofumi Matsuda, Haoshen Zhou, and Itaru Honma. 2006. "Ultrasound-Triggered Smart Drug Release from a Poly(Dimethylsiloxane)-Mesoporous Silica Composite." *Advanced Materials* 18 (23):3083–88.

Kim, Hyehyeon, Saehee Kim, Chiyoung Park, Hyemi Lee, Heon Joo Park, and Chulhee Kim. 2010. "Glutathione-Induced Intracellular Release of Guests from Mesoporous Silica Nanocontainers with Cyclodextrin Gatekeepers." *Advanced Materials* 22 (38):4280–83.

Kim, Kimoon. 2002. "Mechanically Interlocked Molecules Incorporating Cucurbituril and Their Supramolecular Assemblies." *Chemical Society Reviews* 31 (2):96–107.

Knezevic, Nikola Z., Brian G. Trewyn, and Victor S. Y. Lin. 2011. "Functionalized Mesoporous Silica Nanoparticle-Based Visible Light Responsive Controlled Release Delivery System." *Chemical Communications* 47 (10):2817–19.

Lagona, Jason, Pritam Mukhopadhyay, Sriparna Chakrabarti, and Lyle Isaacs. 2005. "The Cucurbit[n]uril Family." *Angewandte Chemie International Edition* 44 (31):4844–70.

Lai, Cheng-Yu, Brian G. Trewyn, Dusan M. Jeftinija, Ksenija Jeftinija, Shu Xu, Srdija Jeftinija, and Victor S. Y. Lin. 2003. "A Mesoporous Silica Nanosphere-Based Carrier System with Chemically Removable CDs Nanoparticle Caps for Stimuli-Responsive Controlled Release of Neurotransmitters and Drug Molecules." *Journal of the American Chemical Society* 125 (15):4451–59.

Lai, Jinping, Xue Mu, Yunyan Xu, Xiaoli Wu, Chuanliu Wu, Chong Li, Jianbin Chen, and Yibing Zhao. 2010. "Light-Responsive Nanogated Ensemble Based on Polymer Grafted Mesoporous Silica Hybrid Nanoparticles." *Chemical Communications* 46 (39):7370–72.

Lee, Jae Wook, S. Samal, N. Selvapalam, Hee-Joon Kim, and Kimoon Kim. 2003. "Cucurbituril Homologues and Derivatives: New Opportunities in Supramolecular Chemistry." *Accounts of Chemical Research* 36 (8):621–30.

Li, Zongxi, Jonathan C. Barnes, Aleksandr Bosoy, J. Fraser Stoddart, and Jeffrey I. Zink. 2012. "Mesoporous Silica Nanoparticles in Biomedical Applications." *Chemical Society Reviews* 41 (7):2590–2605.

Liong, Monty, Sarah Angelos, Eunshil Choi, Kaushik Patel, J. Fraser Stoddart, and Jeffrey I. Zink. 2009. "Mesostructured Multifunctional Nanoparticles for Imaging and Drug Delivery." *Journal of Materials Chemistry* 19 (35):6251–57.

Liong, Monty, Jie Lu, Michael Kovochich, Tian Xia, Stefan G. Ruehm, Andre E. Nel, Fuyuhiko Tamanoi, and Jeffrey I. Zink. 2008. "Multifunctional Inorganic Nanoparticles for Imaging, Targeting, and Drug Delivery." *ACS Nano* 2 (5):889–96.

Liu, Juewen, Xingmao Jiang, Carlee Ashley, and C. Jeffrey Brinker. 2009. "Electrostatically Mediated Liposome Fusion and Lipid Exchange with a Nanoparticle-Supported Bilayer for Control of Surface Charge, Drug Containment, and Delivery." *Journal of the American Chemical Society* 131 (22):7567–69.

Liu, Rui, Puhong Liao, Jikai Liu, and Pingyun Feng. 2011. "Responsive Polymer-Coated Mesoporous Silica as a pH-Sensitive Nanocarrier for Controlled Release." *Langmuir* 27 (6):3095–99.

Liu, Rui, Ying Zhang, and Pingyun Feng. 2009. "Multiresponsive Supramolecular Nanogated Ensembles." *Journal of the American Chemical Society* 131 (42):15128–29.

Liu, Rui, Ying Zhang, Xiang Zhao, Arun Agarwal, Leonard J. Mueller, and Pingyun Feng. 2010. "pH-Responsive Nanogated Ensemble Based on Gold-Capped Mesoporous Silica through an Acid-Labile Acetal Linker." *Journal of the American Chemical Society* 132 (5):1500–1501.

Liu, Rui, Xiang Zhao, Tao Wu, and Pingyun Feng. 2008. "Tunable Redox-Responsive Hybrid Nanogated Ensembles." *Journal of the American Chemical Society* 130 (44):14418–19.

Liu, Simin, Christian Ruspic, Pritam Mukhopadhyay, Sriparna Chakrabarti, Peter Y. Zavalij, and Lyle Isaacs. 2005. "The Cucurbit[n]uril Family: Prime Components for self-Sorting Systems." *Journal of the American Chemical Society* 127 (45):15959–67.

Lu, Jie, Eunshil Choi, Fuyuhiko Tamanoi, and Jeffrey I. Zink. 2008. "Light-Activated Nanoimpeller-Controlled Drug Release in Cancer Cells." *Small* 4 (4):421–26.

Luo, Zhong, Kaiyong Cai, Yan Hu, Li Zhao, Peng Liu, Lin Duan, and Weihu Yang. 2011. "Mesoporous Silica Nanoparticles End-Capped with Collagen: Redox-Responsive Nanoreservoirs for Targeted Drug Delivery." *Angewandte Chemie International Edition* 50 (3):640–43.

Mairal, Teresa, Veli Cengiz Özalp, Pablo Lozano Sánchez, Mònica Mir, Ioanis Katakis, and Ciara O'Sullivan. 2008. "Aptamers: Molecular Tools for Analytical Applications." *Analytical and Bioanalytical Chemistry* 390 (4):989–1007.

Mal, Nawal Kishor, Masahiro Fujiwara, and Yuko Tanaka. 2003. "Photocontrolled Reversible Release of Guest Molecules from Coumarin-Modified Mesoporous Silica." *Nature* 421 (6921):350–53.

Mal, Nawal Kishor, Masahiro Fujiwara, Yuko Tanaka, Takahisa Taguchi, and Masahiko Matsukata. 2003. "Photo-Switched Storage and Release of Guest Molecules in the Pore Void of Coumarin-Modified MCM-41." *Chemistry of Materials* 15 (17):3385–94.

Manzano, Miguel, Montserrat Colilla, and María Vallet-Regí. 2009. "Drug Delivery from Ordered Mesoporous Matrices." *Expert Opinion on Drug Delivery* 6 (12):1383–1400.

Manzano, Miguel, and María Vallet-Regí. 2010. "New Developments in Ordered Mesoporous Materials for Drug Delivery." *Journal of Materials Chemistry* 20 (27):5593–5604.

Meng, Huan, Min Xue, Tian Xia, Yan-Li Zhao, Fuyuhiko Tamanoi, J. Fraser Stoddart, Jeffrey I. Zink, and Andre E. Nel. 2010. "Autonomous In Vitro Anticancer Drug Release from Mesoporous Silica Nanoparticles by pH-Sensitive Nanovalves." *Journal of the American Chemical Society* 132 (36):12690–97.

Nagase, Kenichi, Jun Kobayashi, Akihiko Kikuchi, Yoshikatsu Akiyama, Hideko Kanazawa, and Teruo Okano. 2007. "Effects of Graft Densities and Chain Lengths on Separation of Bioactive Compounds by Nanolayered Thermoresponsive Polymer Brush Surfaces." *Langmuir* 24 (2):511–17.

Özalp, Veli C., and Thomas Schäfer. 2011. "Aptamer-Based Switchable Nanovalves for Stimuli-Responsive Drug Delivery." *Chemistry: A European Journal* 17 (36):9893–96.

Özalp, Veli Cengiz, Fusun Eyidogan, and Huseyin Avni Oktem. 2011. "Aptamer-Gated Nanoparticles for Smart Drug Delivery." *Pharmaceuticals* 4 (8):1137–57.

Park, Chiyoung, Hyehyeon Kim, Saehee Kim, and Chulhee Kim. 2009. "Enzyme Responsive Nanocontainers with Cyclodextrin Gatekeepers and Synergistic Effects in Release of Guests." *Journal of the American Chemical Society* 131 (46):16614–15.

Park, Chiyoung, Kyuho Lee, and Chulhee Kim. 2009. "Photoresponsive Cyclodextrin-Covered Nanocontainers and Their Sol-Gel Transition Induced by Molecular Recognition." *Angewandte Chemie International Edition* 48 (7):1275–78.

Park, Chiyoung, Kyungho Oh, Sang Cheon Lee, and Chulhee Kim. 2007. "Controlled Release of Guest Molecules from Mesoporous Silica Particles Based on a pH-Responsive Polypseudorotaxane Motif." *Angewandte Chemie International Edition* 46 (9):1455–57.

Park, Jun-Hwan, Young-Ho Lee, and Seong-Geun Oh. 2007. "Preparation of Thermosensitive PNIPAM-Grafted Mesoporous Silica Particles." *Macromolecular Chemistry and Physics* 208 (22):2419–27.

Patel, Kaushik, Sarah Angelos, William R. Dichtel, Ali Coskun, Ying-Wei Yang, Jeffrey I. Zink, and J. Fraser Stoddart. 2008. "Enzyme-Responsive Snap-Top Covered Silica Nanocontainers." *Journal of the American Chemical Society* 130 (8):2382–83.

Ruiz-Hernández, Eduardo, Alejandro Baeza, and María Vallet-Regí. 2011. "Smart Drug Delivery through DNA/Magnetic Nanoparticle Gates." *ACS Nano* 5 (2):1259–66.

Saha, Souravi, Ken C-F. Leung, Thoi D. Nguyen, J. Fraser Stoddart, and Jeffrey I. Zink. 2007. "Nanovalves" [Cover art]. *Advanced Functional Materials* 17 (5).

Sauer, Anna M., Axel Schlossbauer, Nadia Ruthardt, Valentina Cauda, Thomas Bein, and Christoph Bräuchle. 2010. "Role of Endosomal Escape for Disulfide-Based Drug Delivery from Colloidal Mesoporous Silica Evaluated by Live-Cell Imaging." *Nano Letters* 10 (9):3684–91.

Schlossbauer, Axel, Johann Kecht, and Thomas Bein. 2009. "Biotin-Avidin as a Protease-Responsive Cap System for Controlled Guest Release from Colloidal Mesoporous Silica." *Angewandte Chemie International Edition* 48 (17):3092–95.

Schlossbauer, Axel, Simon Warncke, Philipp M. E. Gramlich, Johann Kecht, Antonio Manetto, Thomas Carell, and Thomas Bein. 2010. "A Programmable DNA-Based Molecular Valve for Colloidal Mesoporous Silica." *Angewandte Chemie International Edition* 49 (28):4734–37.

Shamah, Steven M., Judith M. Healy, and Sharon T. Cload. 2008. "Complex Target SELEX." *Accounts of Chemical Research* 41 (1):130–38.

Shi, Jinjun, Alexander R. Votruba, Omid C. Farokhzad, and Robert Langer. 2010. "Nanotechnology in Drug Delivery and Tissue Engineering: From Discovery to Applications." *Nano Letters* 10 (9):3223–30.

Singh, Neetu, Amrita Karambelkar, Luo Gu, Kevin Lin, Jordan S. Miller, Christopher S. Chen, Michael J. Sailor, and Sangeeta N. Bhatia. 2011. "Bioresponsive Mesoporous Silica Nanoparticles for Triggered Drug Release." *Journal of the American Chemical Society* 133 (49):19582–85.

Slowing, Igor I., Juan L. Vivero-Escoto, Chia-Wen Wu, and Victor S. Y. Lin. 2008. "Mesoporous Silica Nanoparticles as Controlled Release Drug Delivery and Gene Transfection Carriers." *Advanced Drug Delivery Reviews* 60 (11):1278–88.

Sun, Jiao-Tong, Chun-Yan Hong, and Cai-Yuan Pan. 2010. "Fabrication of PDEAEMA-Coated Mesoporous Silica Nanoparticles and pH-Responsive Controlled Release." *The Journal of Physical Chemistry C* 114 (29):12481–86.

Torney, Francois, Brian G. Trewyn, Victor S. Y. Lin, and Kan Wang. 2007. "Mesoporous Silica Nanoparticles Deliver DNA and Chemicals into Plants." *Nature Nanotechnology* 2 (5):295–300.

Trewyn, Brian G., Supratim Giri, Igor I. Slowing, and Victor S. Y. Lin. 2007. "Mesoporous Silica Nanoparticle Based Controlled Release, Drug Delivery, and Biosensor Systems." *Chemical Communications* (31):3236–45.

Tuerk, C., and L. Gold. 1990. "Systematic Evolution of Ligands by Exponential Enrichment: RNA Ligands to Bacteriophage T4 DNA Polymerase." *Science* 249 (4968):505–10.

Vallet-Regí, María, Francisco Balas, and Daniel Arcos. 2007. "Mesoporous Materials for Drug Delivery." *Angewandte Chemie International Edition* 46 (40):7548–58.

Vallet-Regí, María, Montserrat Colilla, and Blanca Gonzalez. 2011. "Medical Applications of Organic-Inorganic Hybrid Materials within the Field of Silica-Based Bioceramics." *Chemical Society Reviews* 40 (2):596–607.

Vallet-Regí, María, and Eduardo Ruiz-Hernández. 2011. "Bioceramics: From Bone Regeneration to Cancer Nanomedicine." *Advanced Materials* 23 (44):5177–5218.

Vallet-Regí, María, Eduardo Ruiz-Hernández, Blanca González, and Alejandro Baeza. 2011. "Design of Smart Nanomaterials for Drug and Gene Delivery." *Journal of Biomaterials and Tissue Engineering* 1 (1):6–29.

Vasir, Jaspreet K., Maram K. Reddy, and Vinod D. Labhasetwar. 2005. "Nanosystems in Drug Targeting: Opportunities and Challenges." *Current Nanoscience* 1 (1):47–64.

Vivero-Escoto, Juan L., Igor I. Slowing, Chian-Wen Wu, and Victor S. Y. Lin. 2009. "Photoinduced Intracellular Controlled Release Drug Delivery in Human Cells by Gold-Capped Mesoporous Silica Nanosphere." *Journal of the American Chemical Society* 131 (10):3462–63.

You, Ye-Zi, Kennedy K. Kalebaila, Stephanie L. Brock, and David Oupický. 2008. "Temperature-Controlled Uptake and Release in PNIPAM-Modified Porous Silica Nanoparticles." *Chemistry of Materials* 20 (10):3354–59.

Yuan, Li, Qianqian Tang, Dong Yang, Jin Zhong Zhang, Fayong Zhang, and Jianhua Hu. 2011. "Preparation of pH-Responsive Mesoporous Silica Nanoparticles and Their Application in Controlled Drug Delivery." *The Journal of Physical Chemistry C* 115 (20):9926–32.

Zhao, Yan-Li, Zongxi Li, Sanaz Kabehie, Youssry Y. Botros, J. Fraser Stoddart, and Jeffrey I. Zink. 2010. "pH-Operated Nanopistons on the Surfaces of Mesoporous Silica Nanoparticles." *Journal of the American Chemical Society* 132 (37):13016–25.

Zhu, Chun-Ling, Chun-Hua Lu, Xue-Yuan Song, Huang-Hao Yang, and Xiao-Ru Wang. 2011. "Bioresponsive Controlled Release Using Mesoporous Silica Nanoparticles Capped with Aptamer-Based Molecular Gate." *Journal of the American Chemical Society* 133 (5):1278–81.

Zhu, Shenmin, Zhengyang Zhou, Di Zhang, Chan Jin, and Zhiqiang Li. 2007. "Design and Synthesis of Delivery System Based on SBA-15 with Magnetic Particles Formed in Situ and Thermo-Sensitive PNIPA as Controlled Switch." *Microporous and Mesoporous Materials* 106 (1–3):56–61.

Zhu, Yingchun, and Masahiro Fujiwara. 2007. "Installing Dynamic Molecular Photomechanics in Mesopores: A Multifunctional Controlled-Release Nanosystem." *Angewandte Chemie International Edition* 46 (13):2241–44.

Zhu, Yufang, Stefan Kaskel, Toshiyuki Ikoma, and Nobutaka Hanagata. 2009. "Magnetic SBA-15/poly(N-isopropylacrylamide) Composite: Preparation, Characterization and Temperature-Responsive Drug Release Property. *Microporous and Mesoporous Materials* 123 (1–3):107–12.

Zintchenko, Arkadi, Manfred Ogris, and Ernst Wagner. 2006. "Temperature Dependent Gene Expression Induced by PNIPAM-Based Copolymers: Potential of Hyperthermia in Gene Transfer." *Bioconjugate Chemistry* 17 (3):766–72.

4 Mesoporous Silica Nanoparticles in Nanomedicine

Significant progress has been made in understanding the main biological processes that underlie many diseases. However, it remains a major challenge for the biomedical scientific community to reach comparable achievements in the detection, diagnosis, and treatment of these diseases. The main shortcoming lies in the fact that most clinical therapeutic or imaging agents do not efficiently accumulate in the target cell and/or tissue due to their unspecific distribution throughout the body (Shi et al. 2010). As a consequence, conventional therapeutic and imaging agents require high doses and result in significant side effects (Ferrari 2005). Nanomaterials open up the potential to overcome many of these issues and have received noteworthy attention as platforms for therapeutic and imaging applications (K. Cho et al. 2008; Davis, Chen, and Shin 2008; Farokhzad and Langer 2009; Peer et al. 2007). Nanoparticulate platforms exhibit remarkable advantages compared to conventional materials including tunable size, high agent-loading tailored surface properties, controllable or stimuli-responsive drug release kinetics, improved pharmacokinetics, and biocompatibility (Shi et al. 2010; Alexis et al. 2008). Nanoparticles can be specifically targeted to certain regions of the body (i.e., unhealthy cells/tissues) by conjugation with targeting ligands. They can also be designed to contain multiple agents, such as imaging and therapeutic agents, for real-time monitoring of the drug uptake and/or therapeutic responses.

An ideal nanomedical multifunctional nanoplatform must fulfill important requirements. The nanocarrier material must (1) be biocompatible; (2) have high loading and protection capability of desired therapeutic or imaging molecules; (3) have "zero premature release" of drug molecules before reaching its target (this is particularly important in oncology, due to the high toxicity of chemotherapeutic agents); (4) have cell type or tissue specificity and site-targeting ability; (5) have efficient cellular uptake; (6) have effective endosomal escape; and (7) have controllable rate of release to achieve an effective local concentration.

The nanoparticle platforms that have been extensively explored for biomedical applications are organic and inorganic materials. Organic nanoparticles include dendrimers (Khandare et al. 2012), liposomes (Torchilin 2005; Mulder et al. 2009), polymers (Haag and Kratz 2006; Oh et al. 2007), and virus-like particles (Ma, Nolte, and Cornelissen 2012). Inorganic nanoparticles, such as gold (Ghosh et al. 2008), semiconductor nanocrystals (Smith et al. 2008), superparamagnetic nanoparticles (Sun, Lee, and Zhang 2008), and silicon- (Anglin et al. 2008) and silica-based materials (Barbé et al. 2004; Piao et al. 2008), also have been evaluated for biomedical applications. Inorganic nanoparticles are receiving growing attention due to their increased mechanical strength, chemical stability, biocompatibility, and resistance to microbial attack as compared to their organic equivalents (Avnir et al. 2006; Tan et al. 2004; Barbé et al. 2004). In addition, the ceramic matrix efficiently protects entrapped guest molecules against enzymatic degradation or denaturation induced by pH and temperature as no swelling or porosity changes take place in response to variations in the surrounding medium.

Among the different types of inorganic nanomaterials, mesoporous silica nanoparticles (MSNPs) have emerged as promising multifunctional platforms for nanomedicine. Since their introduction in the drug delivery landscape in 2001 (Vallet-Regí et al. 2001), MSNPs are receiving growing scientific interest for their potential applications in the biotechnology and nanomedicine fields (Ambrogio et al. 2011; Ashley et al. 2011; Z. Li et al. 2012; J. Liu et al. 2009; Rosenholm, Sahlgren, and Lindén 2010; Vallet-Regí et al. 2011; Vallet-Regí, Balas, and Arcos 2007; Vallet-Regí, Colilla, and Gonzalez 2011; Vallet-Regí and Ruiz-Hernández 2011; Vivero-Escoto et al. 2010; Wu, Hung, and Mou 2011; Zhao et al. 2010).

The synthesis of MSNPs for biomedical applications needs to satisfy two conditions: (1) well-controlled nucleation and growth rate of MSNPs to produce uniform sizes in the range of 50–300 nm, and (2) nonstickiness of the MSNPs during workup. Broadly speaking, there are two main approaches to synthesizing MSNPs that fulfill these conditions. The first one is the so-called modified Stöber method, which consists of the condensation of silica under basic medium in the presence of cationic surfactants as structure-directing agents. Alkaline and highly diluted conditions are usually used to lead to a negatively charged and more fully condensed surface, to avoid interparticle aggregations (Grün, Lauer, and Unger 1997). The second strategy is the aerosol-assisted synthesis, which allows using not only cationic but also anionic and nonionic surfactants to obtain MSNPs (Arcos et al. 2009; Boissiere et al. 2011; Brinker et al. 1999; Colilla et al. 2010; Y. Lu et al. 1999). The final step in both approaches is the surfactant removal, which normally leads to materials with cylindrical mesopores arranged in a two-dimensional hexagonal fashion, characteristic of MCM-41-type materials (Kresge et al. 1992). These MSNPs exhibit distinctive and advantageous textural and structural features such as high surface area (ca. 1000 m^2/g), high pore volume (ca. 1 cm^3/g), stable mesostructure, tunable pore diameter (2–10 nm), two functional surfaces (exterior particle and inner pore faces), and tunable particle size. The nanoparticle size in the range of interest for targeted intracellular delivery applications falls into the 50–300 nm range, since larger particles cannot easily bypass physical membranes in the body, and smaller MSNPs are difficult to synthesize due to their inherent mesoporosity. As the particle size has been demonstrated to play a pivotal role in the nanoparticle distribution and behavior in living systems (Farokhzad and Langer 2009; Y.-S. Lin and Haynes 2010; Suh, Suh, and Stucky 2009), narrow particle size distributions are preferred. Therefore wet chemical synthesis methods are generally more beneficial than other physical methods such as spray drying, which commonly lead to a relatively wide particle size distribution with nonnegligible fractions of fine and larger particles.

MSNPs possess well-defined structure and high density of surface silanol groups, which can be modified with a wide range of organic functional groups (Bruhwïler 2010; Hoffmann et al. 2006; Hoffmann and Froba 2011). The surface functional groups can play several roles in MSNPs for nanomedicine: (1) to control the surface charge of MSNPs; (2) to chemically link with functional molecules inside or outside the pores; and (3) to control the size of pore entrance for entrapping molecules in the nanopores.

The three methods of surface functionalization for MSNPs are co-condensation, post-synthesis grafting, and surfactant displacement. The co-condensation or one-pot method involves adding the organosilanes directly in the synthesis step together with the silica source (Arcos et al. 2009; Brinker et al. 1999; Bruhwïler 2010; Hoffmann et al. 2006; Hoffmann and Froba 2011). Then the surfactant is removed by ion exchange with an ethanolic solution of ammonium nitrate (Lang and Tuel 2004) or HCl (Burleigh et al. 2001; Huh et al. 2003). The main advantages of this method are a simple operation and homogeneous distribution of functional groups. However, under some conditions the extraction of surfactant may not be complete, depending on the solvent used. The grafting or post-synthesis method consists of incorporating the functional groups once the surfactant removal, by calcination or extraction, has been performed (Lim and Stein 1999). This method brings up many possibilities for placing the different functional groups and also permits the grafting of chemically delicate organic functionalities susceptible to hydrolysis and elimination reactions (Zapilko et al. 2006). One of the main concerns of the grafting method is that sometimes there is a lack of uniformity in the distribution of functional groups if some of them block the mesopore entrances (Ritter and Brühwiler 2009). Finally, the surfactant displacement method consists of the direct surface silylation with simultaneous surfactant extraction without prior calcination by using acidic alcohol as the solvent (H.-P. Lin et al. 2000; Y.-H. Liu, Lin, and Mou 2004). This method leads to uniform monolayer coverage, where the amount of functionalization organosilanes present on the surface can be accurately modulated.

The unique topology of MSNPs provides them with three well-defined domains that can be distinctively functionalized: the silica framework, the nanopores, and the outermost surface. These paramount attributes make MSNPs ideal nanoplatforms to incorporate different multifunctionalities for therapy and diagnostics of several diseases (Figure 4.1) (Ambrogio et al. 2011; Ashley et al. 2011; Rosenholm, Sahlgren, and Lindén 2010; Vallet-Regí, Balas, and Arcos 2007; Vallet-Regí and Ruiz-Hernández 2011; Wu, Hung, and Mou 2011).

The silica framework can incorporate organic molecules as optical imaging (OI) or magnetic resonance imaging (MRI) contrast agents. Magnetic nanoparticles can be also incorporated within the silica matrix to act as MRI contrast agents or as thermoseeds for hyperthermia treatment of cancer. The mesoporous cavities can host and protect a wide variety of organic molecules, such as drugs, proteins, nucleic acids, MRI contrast agents, or photosensitizers for photodynamic therapy (PDT). These moieties can be just adsorbed or covalently linked to the inner mesopore walls. Molecular nanogates can be covalently linked to the pore outlets, blocking the mesopores and preventing the premature release of entrapped molecules. The stimulus-responsive aperture of the nanogates would trigger the cargo release in the place where needed.

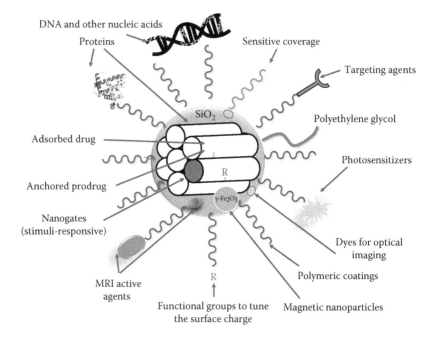

FIGURE 4.1   MSNPs as nanomedical multifunctional nanoplatforms.

The outermost surface of MSNPs can be decorated with targeting ligands for specific cellular uptake, functional groups to tune surface charge, polymeric coatings such as polyethylene glycol (PEG), or stimuli-responsive polymers, and MRI contrast agents, photosensitizers, nucleic acids, or proteins can be also immobilized.

This chapter describes the recent scientific focused on designing MSNPs for nanomedicine. We also tackle the open challenges that need to be addressed that permit the transfer of MSNPs from bench to bedside.

4.1   DRUG DELIVERY

The high intrinsic specific surface areas and pore volumes of MSNPs permit the carrying of a large amount of drug, sometimes exceeding 30% wt. Drug loading can be judiciously tuned, which permits using appropriate nanoparticle concentrations to reach therapeutic levels in vitro and in vivo while minimizing particle-induced toxicity.

The drug loading into MSNPs is usually carried out as the final step, i.e., once the MSNPs have been synthesized. The distinction between material synthesis and drug incorporation permits the independent optimization of the structural and physicochemical properties of the carrier (host) and the conditions for the drug (guest) loading (Rosenholm, Sahlgren, and Lindén 2010). The incorporation of drugs into the MSNPs can be achieved using different strategies. The chosen loading method depends on the molecular structure and the size of the drug relating to the mesopore diameter, its water solubility and cytotoxicity, and the chemical stability of the drug.

There are three main methods to carry out drug loading: physical adsorption from solution into the mesopores, physical adsorption from solution onto the outermost surface of MSNPs, and covalent linking. In addition, other loading methods, such as impregnation by incipient wetness or loading from a drug melt, have been tested.

One of the most remarkable advantages of MSNPs relies on their capability to host hydrophobic drugs, which may be difficult when using other nanocarriers. This is due to the fact that the drug loading is usually carried out as the final step, i.e., once the MSNPs have been synthesized and their surface chemistry has been designed by incorporating suitable functionalities (organic functions, targeting ligands, imaging contrast agents, polymeric coatings, etc.). MSNPs preserve their structural integrity in contact with organic solvents, thanks to the ceramic matrix. For this reason, the drug-loading process can be carried out in nonaqueous media, and repulsive forces between the host (mesoporous matrix) and the guest (drug) are minimized (Rosenholm and Lindén 2008; Rosenholm et al. 2011). The ideal solvent is selected based on drug solubility to make prevailing the host-guest (matrix-drug) interactions over solvent-drug interactions. In this case, electrostatic interactions are not relevant and the driving forces for drug adsorption are mainly hydrogen bonding or polar interactions. It is also possible to promote the host-guest interactions by incorporating functional groups onto the mesoporous matrix that show high affinity for the drug, such as amino groups in the case of adsorption of drugs bearing carboxylic acids. Therefore when dealing with hydrophobic drugs an organic solvent is usually chosen and normally a monolayer, which can be modeled using a Langmuir adsorption isotherm, is observed (Andersson et al. 2004). This permits high control over the loading process and thus the drug-loading degree can be modulated to meet clinical needs.

For more hydrophilic drugs, tailoring the pH of aqueous solvent can be used to reach higher drug-loading levels than those possible from organic solvents (Rosenholm and Lindén 2008). In the mesoporous silica surface, there are plenty of silanol groups that can be present in both a protonated and a deprotonated form in aqueous media depending on the pH (Tourne-Peteilh et al. 2003). As pH at which the net surface of silica is zero (zero charge point) is ca. 2–3, in the absence of specific ion adsorption the silica surface is negatively charged under biologically relevant conditions. Thus electrostatic adsorption of positively charged drugs is a good method for incorporating water-soluble drugs into MSNPs. However, the electrostatic adsorption can be increased by incorporating functional groups, such as weak acids and bases like carboxylic acids or amines, onto the silica surface. Organic functionalization will permit a fine-tuning of effective surface charge under given pH conditions, bringing up the possibility for additional specific host-guest interactions. It is worth mentioning that modulation of surface charge at a given pH depends on both the chemical nature of the incorporated functional groups and the functionalization degree.

It is also possible to covalently link a given drug to diverse functional groups present on the mesopore walls (Rosenholm et al. 2010; Mortera et al. 2009; Tourne-Peteilh et al. 2003), which is especially attractive to prevent premature drug release. However, the drug activity should be preserved after decoupling from the pore wall, which involves the original functional group of the drug molecule used for the covalent grafting re-forming upon detachment. Recent advances have been achieved to covalently link drugs or prodrugs by stimuli-sensitive functional groups such as the pH-hydrolysable hydrazone bond (I. Huang, Sun et al. 2011; Jianquan et al. 2011; C.-H. Lee et al. 2010). Hydrolysis of the pH-sensitive hydrazone bond in the acidic environment of endosomes/lysosomes releases the drug intracellularly from the MSNPs' nanochannels.

Finally, the most widely explored strategy to prevent premature drug release consists of incorporating nanogates onto the outer particle surface to create gate-keeping properties. These nanocaps keep the pores closed to avoid drug loss and open upon the application of a given stimulus, such as changes in pH or temperature; redox potential; application of light, magnetic fields, or ultrasounds; the presence of enzymes, antigens, or aptamer targets; or even dual stimuli (Ambrogio et al. 2011; Vallet-Regí, Balas, and Arcos 2007; Saha et al. 2007; Slowing et al. 2008; Coti et al. 2009). For more detailed information and further discussion on this topic, we recommend that the reader consult Chapter 3 of this book.

4.2 TARGETING

Cell targeting is a major concern in cases such as cancer therapies, where the lack of specificity of antitumor drugs used in conventional therapies results in serious adverse side effects and limited effectiveness due to unspecific action healthy cells. In this sense, much research effort is being dedicated to develop MSNPs with cell-targeting capability to specifically release drugs to unhealthy cells. Targeted delivery to tumors can be achieved by either passive targeting, such as the enhanced permeability and retention effect (EPR), or active targeting of receptors, i.e., surface proteins specifically expressed on the surface of cancer cells, or by a combination of both (Figure 4.2) (Farokhzad and Langer 2009; Langer 1998; Danhier, Feron, and Préat 2010).

4.2.1 PASSIVE TARGETING AND "STEALTH" PROPERTIES

Due to the leaky vasculature and poorly operational lymph system of tumors, MSNPs can exit the blood vessels and accumulate at the tumor site by passive targeting via the enhanced permeability and retention (EPR) effect (Figure 4.2) (Muggia 1999; Maeda, Bharate, and Daruwalla 2009; Torchilin 2011). Tumor tissue accumulation is a passive process requiring a long-circulating half-life to facilitate time-dependent extravasations of drug delivery systems through the leaky tumor microvasculature and accumulation of drugs in the tumor tissue (Jain 1998). Therefore even in the absence of targeting ligands, drug delivery systems can be engineered to better target a particular tissue, or nonspecifically absorbed by cells, by optimizing their biophysicochemical properties (Verma et al. 2008; Gratton et al. 2008). The diffusion rate in the extracellular spaces of the tumor is governed by the size and the surface charge of the particles. Small particles easily diffuse in and out of the tumor vasculature, and the concentration of the drug carrier at the tumor site might be low. Free movement is also affected by surface charge, and negatively charged or weakly positively charged

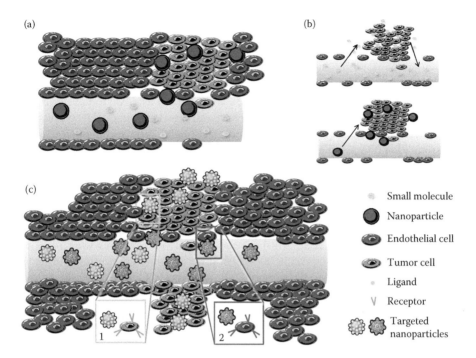

FIGURE 4.2 **(See color insert.)** (a) Passive targeting of nanoparticles, which reach tumor cells selectively through the leaky vasculature surrounding the tumors. (b) Influence of the size in the retention in the tumor tissue. Small molecules, such as drugs, diffuse without restraint in and out of the tumor blood vessels, which prevents their effective accumulation in the tumor. By contrast, the relatively large size of nanoparticles (20–200 nm) prevents their diffusion back into the blood stream, which results in progressive accumulation, i.e., the EPR effect. (c) Strategies for active targeting of nanoparticles. Ligands anchored on the surface of nanoparticles bind to receptors overexpressed by (1) cancer cells or (2) angiogenic endothelial cells. *Sources*: Muggia (1999); Maeda, Bharate, and Daruwalla (2009); Torchilin (2011); Langer (1998).

particles in the 50–150 nm range easily pass through tumor tissue (Davis, Chen, and Shin 2008).

Opsonization and protein association on the surface of a particle will immediately lead to massive liver uptake and toxicity and will also contribute to particle aggregation. Opsonins are proteins, mainly fibrinogen, immunoglobulins, albumin, and complement components, that, after binding to a particle's surface, augment their phagocytosis and uptake via cells of the reticuloendothelial system (RES) (Esmaeili et al. 2008; Sadzuka et al. 2006; Zeisig et al. 1996). This phenomenon is a major concern for targeted therapies because it will decrease the targetability of the nanoparticles and lead to rapid clearance of the drug carriers from the blood circulation. One of the main developed strategies to overcome this drawback consists of covering the surface of the MSNPs with different polymers to obtain "stealth" nanoparticles. Providing MSNPs with "stealth" characteristics promotes EPR effect of drug carriers at the tumor site, and their blood circulation half-lives are prolonged (Harper et al. 1991; Veronese and Pasut 2005).

Polyethylene glycols (PEGs) are the main family of polymers used to obtain "stealth" particles (Pinholt et al. 2011; Morille et al. 2011; Faure et al. 2009; Dosio et al. 2010; Paillard et al. 2010; Vonarbourg et al. 2006; Erbacher et al. 1999). The "stealth" properties of PEGylated particles are generally ascribed to the steric hindrance and repulsion effects of PEG chains against blood proteins and macrophages, which are tightly related to the PEG molecular weight, surface chain density, and conformation (Owens Iii and Peppas 2006; Claesson et al. 1995; Dufort, Sancey, and Coll 2012). One major challenge is to find the optimal PEG molecular weight and chain density on MSNPs to minimize both their binding to blood proteins and their uptake by human macrophages (Pirollo and Chang 2008). He et al. investigated the influence of PEG molecular weight and packing density on the adsorption of serum proteins to MSNPs, whose PEG chains were linked to the MSNPs using silane-coupling chemistry (He et al. 2010). The results derived from this work revealed that PEG molecular weights of 10,000–20,000 were optimum for minimizing non-specific protein adsorption. Furthermore, the optimum PEG molecular weight may be a function of the chemistry used to covalently link PEG to the MSNP surface. Recently, the same research group carried out in vivo assays to evaluate the effect of PEGylation of MSNPs in the in vivo biodistribution and urinary excretion by tail-vein injection in ICR mice (He et al. 2011). The results indicated that PEGylated MSNPs escaped more easily from capture by liver, spleen, and lung tissues, possessed longer blood-circulation lifetime, and were more slowly biodegraded and correspondingly had a lower excreted amount of degradation products in the urine compared to non-PEGylated MSNPs.

4.2.2 Active Targeting

Active targeting not only enhances, through interactions between the carrier and the cancer cells, the retention of the carrier at the tumor site but also facilitates cellular uptake, leading to increased intracellular concentrations of the drug and consequently improved therapeutic efficacy (Figure 4.2) (Danhier, Feron, and Préat 2010; Farokhzad and Langer 2009; Langer 1998). The selectivity is a function of the ability of the nanocarrier to be internalized by the targeted cell population. Hence, different targeting ligands, such as sugars (Gu et al. 2007; Park et al. 2008; Brevet et al. 2009; Hocine et al. 2010; Luo et al. 2011), monoclonal antibodies (Tsai et al. 2009; Kamphuis et al. 2010), DNA aptamers (Zhu et al. 2009; Özalp 2011), or folic acid (FA) (Gu et al. 2007; Rosenholm et al. 2008, 2009, 2010; Liong et al. 2008; Lebret et al. 2008; Wang et al. 2010; J. Lu et al. 2010, 2012), have been attached to the external surface of MSNPs to promote cell specificity. These systems are sustained in the ability of the targeting agents to selectively bind receptors that are overexpressed on the cancer cell surface to trigger receptor-mediated endocytosis (Figure 4.3). Among these possibilities, the functionalization of MSNPs with FA has emerged as an attractive alternative for targeted drug delivery (Low, Henne, and Doorneweerd 2007) since α-folate receptor (FR) is upregulated in several types of human cancer cells, such as ovarian, endometrial, colorectal, breast, lung, renal

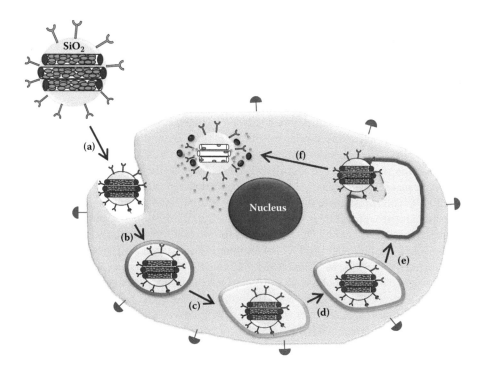

FIGURE 4.3   (See color insert.) Schematic representation of the intracellular trafficking of MSNPs. (a) The cellular uptake takes place through specific (ligand-receptor) and nonspecific (hydrophobic, Coulombic) binding interactions. (b) The internalized MSNPs are delivered to intermediate compartments, such as caveosomes. (c) Such compartments are transported to early endosomes and then to sorting endosomes. From sorting endosomes a fraction of the MSNPs are sorted back to the cell exterior through recycling endosomes (not shown in this figure). (d) The remaining fraction is transported to secondary endosomes, (e) which then fuse with lysosomes. (f) The MSNs escape the endolysosomes and enter the cytosolic compartment. Then the MSNPs can release their cargo under a given change in the surrounding medium, which acts as a stimulus that triggers drug release.

cell carcinoma, brain metastases derived from epithelial cancer, and neuroendocrine carcinoma (Sudimack and Lee 2000; Elnakat and Ratnam 2004). Successful in vitro cell-specific drug delivery to cancer cells using MSNPs functionalized with FA has been achieved. For instance, Rosenholm et al. evaluated the internalization of MSNPs functionalized with FA in cell lines expressing different levels of FR (Rosenholm et al. 2008). Results revealed that five times more particles were internalized by cancer cells expressing FR as compared to the normal cells expressing low levels of such receptor. Besides, not only the number of MSNPs internalized per cell but also the fraction of cells that had internalized MSNPs was higher. The total number of particles internalized by the cancer cells was therefore about one order of magnitude higher than the total number of particles internalized by normal cells, a difference high enough to be of significant biological importance. Moreover, the selective uptake of FA-tagged MSNPs was demonstrated under coculture conditions, where both cancer cells with elevated FR expression and noncancerous cells were incubated together.

Liong et al. (Liong et al. 2008) demonstrated that the cytotoxicity of MSNPs-FA loaded with the antitumor drug camptothecin (CPT) was higher in cell lines over-expressing FR, such as pancreatic PANC-1 cancer cells (60% cell death based on a viability assay), than in control noncancerous cells, such as foreskin fibroblast HFF (30% cell death). Besides, the cytotoxicity between the FA-modified and unmodified drug-loaded MSNPs was similar for the HFF since these cells do not overexpress the receptors. These results demonstrate that FA modification to the MSNPs can increase the particle uptake and deliver more drugs to the cancer cells but not to the noncancerous fibroblast. Similar numbers were reported by Zhu et al. (Zhu et al. 2009), who investigated the effect of doxorubicin (DOX)-loaded MSNPs surface functionalized with aptamer conjugates on cell viability in cancerous HeLa (40% viability) cells and healthy QGY7703 cells (60% viability). Recently, Rosenholm et al. attached methotrexate (MTX), an antitumor drug structurally similar to FA, to the surface of MSNPs. MTX exerted a dual function, acting as targeting ligand and cytotoxic agent (Rosenholm et al. 2010). MTX-loaded MSNPs induced an apoptotic cell death of about 33% after 72 hours in the cancerous HeLa cell line, compared to noncancerous HEK293 cells where no apoptosis over the control was observed.

The research group headed by Zink has provided noteworthy advances into this research arena by carrying out in vivo studies of targeted versus nontargeted MSNPs. These authors reported for the first time that MSNPs were effective for antitumor drug delivery and that the tumor suppression was significant with a subcutaneous human breast cancer xenograft in mice (J. Lu et al. 2010). In this study, the subcutaneous tumors in mice were virtually eliminated by treating with CPT-loaded MSNPs or CPT-loaded FA-modified-MSNPs. More recently, this research group reported the efficacy of MSNPs using two different human pancreatic cancer xenografts on different mouse species. Significant tumor suppression effects were achieved with CPT-loaded MSNPs. Dramatic improvement of the potency of tumor suppression was obtained by surface modifying MSNPs with FA.

4.3　GENE THERAPY

Gene therapy can be defined as the treatment or prevention, in the case of DNA vaccines, of genetic disorders by the transfer of therapeutic genetic material (DNA or RNA) into a group of cells (a tissue or organ) (Feuerbach and Crystal 1996; Orkin 1986; Rubanyi 2001; Labhasetwar 2005; Mancuso et al. 2009). Nowadays, gene therapy offers great opportunities for treating cancer and other acquired diseases, such as cardiovascular and neurodegenerative ones.

The delivery of nucleic acids aims at correcting or modifying the expression of the gene influencing the disease process. This can be achieved by means of replacing deleterious mutant alleles with functional ones, up- or downregulating the expression of target genes, or even inducing the malignant cell apoptosis in the case of cancer. The incorporation of genetic material into a host genome is a process that requires delivery of the nucleic acids into the cell nucleus. Another alternative involves RNA interference, a mechanism of action in which small interfering RNA (siRNA) sequences bind to target messenger RNA and initiate its degradation, leading to gene silencing (Moazed 2009). In this case of RNA targeting, the gene delivery into the nucleus is not generally required.

Two different approaches have been defined during the development of gene therapy. Thus ex vivo gene therapy involves the genetic modification and reinfusion of specific cells isolated and purified from a patient (Hauser et al. 2000). On the other hand, in vivo gene therapy involves direct transfer of therapeutic genetic material into the tissue of the patient using a vector (Klink et al. 2004). A vector can be described as a system able to carry out the following functions: (1) enabling delivery of genes into the target cells and their nucleus; (2) providing protection from gene degradation; and (3) ensuring gene transcription in the cell. The main problem with the clinical application of cancer gene therapy is the lack of gene transfer vectors that are safe, efficacious, and tumor-selective. Two types of vectors have been employed as vehicles for gene transfer, viral and nonviral vectors. Viral vectors for gene transduction, such as retroviruses, adenoviruses, adeno-associated viruses, lentiviruses, and herpes simplex viruses, are highly efficient. However, although viral vectors are designed to eliminate the viral pathogenicity, there are biosafety issues including random recombination, oncogenic potential, and immunogenicity, and limitations regarding scale-up procedures. For this reason, the current research is focused on designing synthetic nanovectors for gene transfection that show higher safety and biocompatibility than viral vectors. Thus cationic polymers (Edinger and Wagner 2011), polymeric nanocapsules (Woodrow et al. 2009), liposomes (Tseng, Mozumdar, and Huang 2009), dendrimers (Mintzer and Simanek 2008), and inorganic nanoparticles (Slowing et al. 2008; Rosi et al. 2006; Sokolova and Epple 2008; De, Ghosh, and Rotello 2008; Gonzalez et al. 2011) have been developed as nonviral vectors.

A nonviral vector suitable for gene delivery must fulfill important requisites to deal with the biological barriers and achieve successful gene transfection (Mintzer and Simanek 2008; Lechardeur, Verkman, and Lukacs 2005):

1. Protecting DNA from degradation by serum nucleases in the blood by compacting DNA into spherical complexes. This is achieved through the complexation of DNA via electrostatic interaction between the negatively charged phosphate backbone of DNA and cationic molecules.
2. Exhibiting positive net surface charge at physiological pH to promote vector wrapping by the negatively charged cell membrane and cellular uptake. Otherwise, internalization or cellular uptake of naked DNA by plasma membrane permeation is hindered by the size and negative charge of the DNA. In addition, vectors containing targeting ligands are internalized more rapidly and efficiently by receptor-mediated endocytosis than nontargeted vectors, which are taken by nonspecific endocytosis.
3. Possessing a mechanism to protect DNA from the acid environment inside endosomes and perform endosomal escape. At this point, macromolecules that have amine groups with low pK_a values display a "proton sponge" behavior. This means that they are capable of buffering the endosomal vesicle, which leads to endosomal swelling and lysis, thus releasing the DNA or

DNA/vector complex into the cytoplasm. Once released into the cytoplasm, DNA or DNA/vector complexes en route to the nucleus must overcome additional barriers in the cytosol that hamper delivery of the DNA into the nucleus of the host cell. The cytoplasm is abundant in nucleases and various organelles that cause degradation and hinder vector movements. The nuclear envelope is the ultimate obstacle to the nuclear entry of DNA. There is a higher transfectability in dividing cells since disassembly of the nuclear membrane assists vector entry into the nuclei during mitotic cell division. In nondividing cells inefficient nuclear transport through the nuclear pore complex limits the efficacy of nonviral vectors. Finally, once DNA reaches the nucleus, it becomes accessible to the transcriptional machinery and the resulting mRNA is at the end translated into the therapeutic protein.

Among inorganic nanoparticles, MSNPs have emerged as potential delivery platforms of genetic material. The design of MSNP-based nonviral vectors requires providing the external silica surface with a positive charge in order to maximize the binding of negatively charged nucleic acids and their delivery into cells. The general method for introducing such cationic charge on the surface of MSNPs is their functionalization with amine polymeric structures (Rosenholm et al. 2008; Gonzalez et al. 2009; Soler-Illia and Azzaroni 2011; Bhattarai et al. 2010).

In 2004, Lin and coworkers developed for the first time a gene transfection system based on MSNPs functionalized with the second generation (G2) of polyamidoamine (PAMAM) dendrimers (Figure 4.4) (Radu, Lai, Jeftinija et al. 2004). PAMAMs are effective gene transfection agents by themselves (Dennig and Duncan 2002; Esfand and Tomalia 2001). However, only those PAMAMs of high generations (G > 5) have been shown to be efficient in gene transfection (Dennig and Duncan 2002). The necessary procedures for the synthesis and purification of these high-generation PAMAMs are usually tedious and the achieved yield is low. By contrast, the low-generation PAMAMs (G < 3) are nontoxic and easy to synthesize. However, the limited surface charges of the low-generation PAMAMs avoid efficient complexation with plasmid DNAs in solution (Dennig and Duncan 2002). Thus the combination of low-generation PAMAMs and MSNPs brings up the possibility of developing hybrid nanomaterials with optimal plasmid DNA complexation (Radu, Lai, Jeftinija et al. 2004). To design such hybrid nanomaterials, isocyanatopropyl-functionalized MSNPs were first loaded with the fluorescent dye Texas Red with the aim of tracking the distribution of the MSNPs once internalized into the cells. Then G2-PAMAM dendrimers were covalently linked through urea linkages to the outer surface of MSNPs, blocking the pore entrances. The main role that the attached dendrimers play consists of efficiently complexing DNA at physiological pH, where the dendrimers are positively charged, and protecting the DNA from enzymatic cleavage. The transfection efficiency of the system was evaluated by using the G2-PAMAM-capped MSNP material to complex a plasmid DNA (pEGFP-C1) that codes for an enhanced green fluorescence protein. The successful delivery of the plasmid to the cell nucleus was confirmed with the observation of significant expression of the green fluorescent protein in the transfected cells (Figure 4.4).

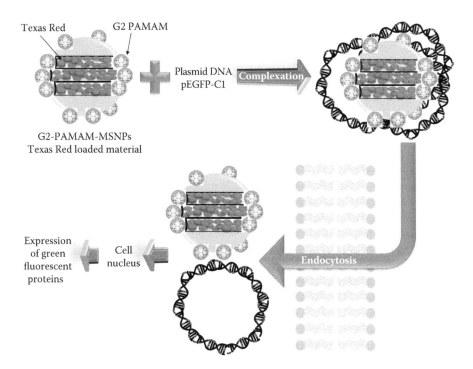

FIGURE 4.4 Schematic representation of a nonviral gene transfection system based on Texas Red-loaded G2-PAMAM dendrimer-capped MSNPs complexed with a plasmid DNA (pEGFP-C1) that codes for an enhanced green fluorescence protein. *Source*: Radu, Lai, Jeftinija et al. (2004).

Polyethyleneimine (PEI)-coated MSNPs have also been used as nucleic acid delivery systems (Xia et al. 2009; Hom et al. 2010). PEIs are synthetic cationic polymers that compact DNA and siRNA into complexes that are effectively taken up in cells to make nucleic acid delivery and gene therapy feasible (Boussif et al. 1995; Godbey, Wu, and Mikos 1999; Urban-Klein et al. 2004). Although PEI itself is used as a delivery vehicle, this polymer can also be attached to nanoparticle surfaces via covalent and electrostatic interactions to achieve the same goal (Anglin et al. 2008; Haag and Kratz 2006; Piao et al. 2008; Sun, Lee, and Zhang 2008; Vivero-Escoto et al. 2010). Complexing PEI with the nanoparticles has the potential advantage of facilitating DNA and siRNA delivery by a multifunctional platform that also allows imaging, targeting, and concurrent drug delivery. PEI polymeric coatings enhance the particle uptake into cells and facilitate endosomal escape for the nucleotide delivery (Duan and Nie 2007). Different studies have demonstrated that although low-molecular weight (MW) PEIs are not cytotoxic, these polymers are ineffective at transfecting nucleotides in contrast to the high-MW PEIs. In this sense, the size (MW), compactness, and chemical modification of the PEI affect the efficacy and toxicity of this polymer (Florea et al. 2002; Neu, Fischer, and Kissel 2005). PEI cytotoxicity takes place via a proton sponge effect, which involves proton sequestration by the polymer surface that leads to heightened proton pump activity inside the cell, osmotic swelling of the endocytic compartment, endosomal rupture, and ultimately cell death by a mitochondrial mediated mechanism (Xia et al. 2006, 2007). Taking into account all these premises, Zink and coworkers investigated the effect of several sizes of PEI polymer coatings on MSNPs regarding cellular uptake, cellular toxicity, and efficiency of drug and nucleic acid delivery (Xia et al. 2009). PEIs ranging from molecular weight of 0.6 to 25 kDa were explored for achieving cellular delivery versus reduction of toxicity. The MSNP's surface was modified with PEI via electrostatic interactions. To achieve this goal the authors first functionalized MSNPs with phosphonate groups and subsequently coated the nanoparticles with PEIs of different MW (0.6, 1.2 1.8, 10, and 25 kDa). Results indicated that the reduction of the polymer size is capable of scaling back the cytotoxic effect and that particles coated with PEIs of 10 kD or less still preserve the attribute of facilitated cellular uptake of cationic nanoparticles, most probably due to high-avidity membrane binding and efficient membrane wrapping, which allow these particles to enter cellular endocytic compartments. In addition it is demonstrated that the cellular uptake of PEI-coated MSNP, regardless of polymer size, is considerably enhanced compared to unmodified MSNP (silanol surface) or particles coated with phosphonate or PEG groups. This work also shows that the PEI coating of the MSNPs does not affect their capability to host and release drug molecules, providing a dual delivery system.

One of the major concerns that limit the use of chemotherapy for cancer therapy is the development of multidrug resistance (MDR) in cancer cells during the treatment (Dubikovskaya et al. 2008; Gottesman, Fojo, and Bates 2002; Pakunlu et al. 2004). MDR can be divided into two distinct classes, pump and nonpump resistance (Pakunlu et al. 2004). Membrane proteins, P-glycoprotein (Pgp) and multidrug resistance-associated protein (MRP), have been shown to be the main players for pump resistance to a broad range of structurally and functionally distinct cytotoxic agents (Gergely et al. 1998). The main mechanism of nonpump resistance is an activation of cellular antiapoptotic defense, mainly by Bcl-2 protein. Most of the anticancer drugs trigger apoptosis and simultaneously activate both pump and nonpump cellular defense of multidrug resistance, which prevents cell death. Therefore to effectively suppress the overall resistance to chemotherapy, it is essential to simultaneously inhibit both pump and nonpump mechanisms of cellular resistance by targeting all the intracellular molecular targets (Pakunlu et al. 2004; Pakunlu, Cook, and Minko 2003; Lima et al. 2004). The field of RNA interference therapeutics has experienced significant advances since the first demonstration of gene knockdown in mammalian cells (Hannon and Rossi 2004). Small interfering (siRNA)-based formulations offer significant potential as therapeutic agents to induce potent, persistent, and specific silencing of a broad range of genetic targets (Whitehead, Langer, and Anderson 2009). Special sequences of siRNAs targeted against messenger-RNA (mRNA) encoding major proteins responsible for pump and nonpump cellular defense have been developed and shown a substantial efficacy in vitro (Y. Wang et al. 2006; Yano et al. 2004). However, reports on delivering such types of siRNA simultaneously with a traditional antitumor drug to cancer cells for enhanced chemotherapy efficacy have been scant, due to the lack of efficient codelivery methods. Thus the possibility of developing MSNPs as codelivery systems of MDR suppressors and antitumor drugs has emerged as an attractive strategy to overcome this issue and enhance the efficacy of chemotherapy (Saad, Garbuzenko, and Minko 2008). Chen et al. proposed MSNPs to simultaneously deliver Bcl-2-targeted siRNA and DOX (Chen et al. 2009). MSNPs were functionalized with G2-PAMAM dendrimers to immobilize siRNA by using the method previously described by Lin and coworkers (Radu, Lai, Jeftinija et al. 2004). Then G2-PAMAM-modified MSNPs were loaded with DOX. Results derived from in vitro assays demonstrated that by delivering DOX and Bcl-2 siRNA simultaneously into ovarian human cancer cells, the Bcl-2 siRNA effectively silenced the Bcl-2 mRNA, significantly suppressed the nonpump resistance, and substantially enhanced the anticancer action of DOX.

Meng et al. tested the utility of PEI-coated MSNPs as a dual-delivery platform of Pgp siRNA and DOX (Meng et al. 2010). MSNPs were functionalized with a phosphonate group, which allowed electrostatic binding of DOX to the porous interior. DOX release was achieved by acidification of the medium under abiotic and biotic conditions. In addition, phosphonate modification also allowed exterior coating with PEI, which endowed the MSNPs the ability to simultaneously deliver Pgp siRNA. The dual delivery of DOX and siRNA in KB-V1 cells was capable of increasing the intracellular as well as intranuclear drug concentration to levels exceeding that of free DOX or the drug being delivered by MSNPs in the absence of siRNA codelivery. These results also demonstrate that it is possible to use MSNPs as smart platforms to effectively deliver an siRNA that knocks down gene expression of a drug exporter, which can be used to improve drug sensitivity to a chemotherapeutic agent.

4.4 IMAGING

Noninvasive imaging techniques are of foremost relevance in clinical diagnostics. Diverse imaging modalities are available, ranging from techniques that permit whole-organism anatomical imaging such as magnetic resonance imaging (MRI) and computed tomography (CT), to those that provide specific molecular imaging such as positron electron tomography (PET) and optical fluorescence. Imaging technologies benefit greatly from using nanoparticle-based contrast agents (Wu, Hung, and Mou 2011; H. S. Choi and Frangioni 2010; E. Cho et al. 2010; Taylor-Pashow et al. 2010; Vivero-Escoto, Huxford-Phillips, and Lin 2012). Nanoparticle-based contrast agents have to fulfill important prerequisites: (1) forming stable colloidal solutions in diverse in vitro and in vivo environments; (2) possessing chemical stability under diverse physiological conditions (solvent, polarity, reducing environment, ionic strength, pH, or temperature); (3) exhibiting limited nonspecific binding to avoid macrophagocytic system uptake; (4) having programmed clearance mechanisms; (5) showing high sensitivity and selectivity for the target (i.e., antigen, cell, or tissue); (6) exhibiting good image contrast, with high signal-to-noise ratio; and (7) having relatively long circulation time in the blood if administered intravenously.

These nanomaterials are designed to allow long-term quantitative imaging at low doses and be safely cleared from the body after imaging is completed. In this sense, MSNPs-based contrast agents are promising platforms to meet the aforementioned prerequisites. The main attractive features of MSNPs for biomedical imaging applications are their well-defined and tunable structures, i.e., size, morphology and porosity, and surface chemistry. And no less important is the fact that MSNPs are effectively "transparent"; i.e., they do not absorb light in the near-infrared (NIR), visible (VIS), and ultraviolet (UV) regions or interfere with magnetic fields.

Silica-based imaging nanoparticles are mostly used for optical imaging (OI), MRI, or a combination of both modalities. OI is a low-cost technology that allows for rapid screening, while MRI can offer high resolution and the capability to simultaneously obtain physiological and anatomical information.

4.4.1 OPTICAL IMAGING

OI is a potent modality in which specific probes are excited by incident light, normally in the visible or NIR regions, and emit light at a lower energy than that of the incident light. Since radiation is scattered and absorbed quickly within the body, the resolution for OI is limited to 1–2 mm. The need for deeper penetration depths for most clinical applications is overturning optical techniques into the NIR region. Nevertheless, the versatility of OI in terms of availability of a variety of contrast agents, avoidance of radiopharmaceuticals, and relatively low cost of required instrumentation make this technique complementary to other modalities such as MRI and PET.

As previously mentioned, MSNPs have emerged as potent nanoplatforms for designing novel OI contrast agents. Thus the incorporation of dye molecules into MSNPs has been accomplished to investigate cellular internalization and cell tracking (Vivero-Escoto et al. 2010; Slowing, Trewyn, and Lin 2006; J. Lu, Liong, Sherman et al. 2007; J. Lu, Liong, Zink et al. 2007; Liong et al. 2009; Y.-S. Lin et al. 2005). Rosenholm et al. developed MSNPs incorporating both fluorescent and targeting moieties (Rosenholm et al. 2008). In a first step MSNPs were first functionalized by surface hyperbranching polymerization of poly(ethyleneimine) (PEI). Then a fluorescent dye, in this case fluorescein isothiocyanate (FITC), was covalently linked to PEI, making MSNPs visible by fluorescence microscopy and flow cytometry. Folic acid (FA), which was also covalently linked to the hyperbranched PEI layer, was used as the targeting ligand, and the cancer-specific internalization of these particles was tested on tumor and healthy cells (Figure 4.5).The presence of FITC allowed the use of flow cytometry to quantify the mean number of nanoparticles internalized per cell. The biospecifically tagged MSNP-PEI-FITC-FA system was shown to be noncytotoxic and able to specifically target folate receptor-expressing cancer cells also under coculture conditions. Later on, the same research group evaluated the intracellular drug delivery ability of this system. They used two hydrophobic fluorophores, DiI (1,1′-dioactadecyl-3,3,3′,3′-tetramethindocarbocyanine perchlorate) and DiO (3,3-dioactadecyloxacarbocyanine perchlorate), as model drugs, which made it possible to follow the intracellular release by confocal fluorescence microscopy. Furthermore, the use of fluorophores as model cargo allowed quantification of the intracellular delivery using flow cytometry. In vitro assays demonstrated that that the nanoparticles were taken up by receptor-mediated endocytosis followed by accumulation in the endosomal compartment and subsequent release of cargo into the interior of the cell. Therefore besides the selectivity of the developed nanoparticles for cancer cells, the incorporated agent was shown to be able to escape from the endosomes into the cytoplasm, which is essential for successful intracellular delivery.

However, it is well known that OI usually suffers from the attenuation of photon propagation in living tissue and poor signal-to-noise ratio due to tissue autofluorescence. To overcome these shortcomings Lo and coworkers developed NIR MSNP-based probes (C.-H. Lee, Cheng et al. 2009). The authors entrapped indocyanine green (ICG) into MSNPs by electrostatic interactions with trimethylammonium groups incorporated into the silica matrix, affording MSNP-TA-ICG. ICG has been approved by the FDA as an optical contrast agent for clinical applications, exhibiting characteristic fluorescent excitation and emission wavelengths ($\lambda_{ex}/\lambda_{em}$ = 800/820 nm) in the NIR window. For this reason the incorporation of ICG molecules into MSNPs is a good approach to designing novel nanodevices for in vivo optical imaging. Thus the homogeneous dispersion of ICG molecules in the large surface area of MSNPs efficiently prevents their aggregation and self-quenching. Besides, MSNPs protect ICG molecules from degradation and decrease the associated immune response. The authors evaluated the biodistribution of this OI probe in rat and mouse models. Optical images evidenced that after intravenous injection the nanoparticles immediately accumulated in the liver, followed by the kidneys, lungs, spleen, and heart. This was the first report of MSNPs functionalized with NIR-ICG capable of optical imaging in vivo. Later on, the same research group evaluated the

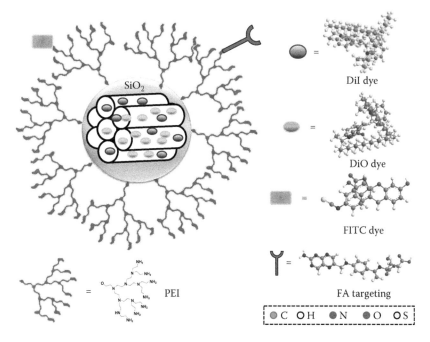

FIGURE 4.5 Schematic representation of MSNPs functionalized by surface hyperbranching polymerization of poly(ethylene imine) (PEI). Further functionalization of PEI with fluorescein isothiocyanate (FITC) and folic acid (FA) for imaging and targeting purposes, respectively. Loading of MSNPs-PEI-FITC-FA system with two hydrophobic fluorophores, DiI (1,1′-dioactadecyl-3,3,3′,3′-tetramethindocarbocyanine perchlorate) and DiO (3,3-dioactadecyoloxacarbocyanine perchlorate), as model drugs, made it possible to follow the intracellular release by confocal fluorescence microscopy. *Source*: Rosenholm et al. (2009).

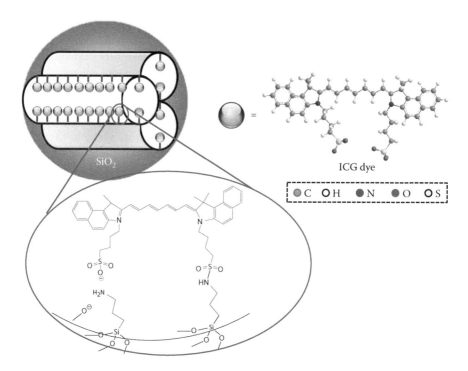

FIGURE 4.6 MSNPs functionalized with aminopropyl groups and containing indocyanine green (ICG) NIR fluorophore immobilized by covalent linkages or ionic bond interactions. *Source*: Lee, Cheng et al. (2009).

role that MSNPs' surface charge played in their biodistribution, clearance from circulation, and excretion (Souris et al. 2010). MSNPs were incorporated with ICG via covalent or ionic bonding, to derive comparable constructs of different net surface charge (Figure 4.6) (C.-H. Lee, Cheng et al. 2009). The results indicated that by tailoring the surface charge of MSNPs, it is possible to control their excretion rates and biodistribution of MSNPs, a functionality that could lead to their widespread clinical use as targetable contrast agents and traceable drug delivery platforms.

4.4.2 Magnetic Resonance Imaging

MRI is a powerful noninvasive diagnostic tool used in medicine that not only provides high-resolution three-dimensional anatomical images of soft tissue but also quantitatively assesses disease pathogenesis by measuring upregulated biomarkers (Terreno et al. 2010). MRI takes advantage of the remarkable range of the physical and chemical properties of water protons (Na, Song, and Hyeon 2009; Villaraza, Bumb, and Brechbiel 2010). The signal detected in MRI results from the combination of total water signal (proton density) and the magnetic properties of the tissues. Concretely, the magnetic properties are the longitudinal relaxation time, T_1, and the transverse relaxation time, T_2, which depend on the physicochemical environment of a given tissue and are modified in the presence of a pathological state.

The reciprocals of these values represent the longitudinal and transverse relaxivities, r_1 and r_2, respectively. Contrast enhancement agents are used in imaging to increase the signal difference between the area of interest and the background. There are two main contrast mechanisms that classify the main types of contrast agents (Khemtong, Kessinger, and Gao 2009). The T_1 agents generate a positive image contrast by increasing the longitudinal relaxation rates of surrounding water protons. The most common T_1 contrast agents are paramagnetic complexes, including mainly chelates of Mn(II), Mn(III), and Gd(III) ions, with Gd-based agents the most common. The T_2 contrast agents generate a negative image contrast by increasing the transverse relaxation rates of water. T_2 contrast agents are mainly superparamagnetic nanoparticles, the most used being iron oxide nanoparticles (IONPs), which can be strongly magnetized under an external magnetic field and lead to a considerable distortion of the local magnetic field.

Unlike T_1 agents, where a chemical exchange between bound and free water molecules is required for the relaxation process, T_2 agents produce much stronger magnetic susceptibility, affecting a larger number of water molecules and thus yielding higher sensitivity of detection.

The development of MRI contrast agents has brought great attention on gadolinium because of its large number of unpaired electrons and relatively long electronic relaxation (Vakil et al. 2009). Recent evolution of molecular imaging has generated a further demand for more sensitive and targeted contrast agents. Nanoparticulate MR contrast agents have also been fabricated based on MSNPs with the aim of taking advantage of their high internal surface areas (Taylor-Pashow et al. 2010). Mesoporous silica is a good carrier for the metal because its high porosity allows water to move freely in and out of the frame; thus the rigidity of the frame hampers the rotational movement of the metal and improves the relaxation of water. Hence, the incorporation of Gd(III) compounds into MSNPs has been proposed as a good approach to designing improved MRI contrast agents.

Lin et al. reported the incorporation of Gd(III) onto MSNPs by loading mesoporous channels with $GdCl_3 \cdot 6H_2O$ (Y.-S. Lin et al. 2004). This system displayed much higher relaxivities, r_1 and r_2, than the commercial complex Magnevist [Gd(DTPA) $(H_2O)]^{2-}$ (DTPA = diethylenetriaminepentaacetic acid). The authors suggested that this was likely because more than one water molecule could be coordinated to metal, leading to a substantial decrease of the Gd metal ion rotation motion in the framework.

Another strategy consisted of synthesizing MSNPs entrapping Gd_2O_3 clusters inside their pores ($Gd_2O_3 \cdot$MCM-41) following a one-step method (Figure 4.7, A) (S. A. Li, Liu et al. 2011). The performance of $Gd_2O_3 \cdot$MCM-41 as a qualified MRI contrast agent was tested in vivo in mice. The $Gd_2O_3 \cdot$MCM-41 NPs led to considerable enhancement of in vivo T_1-weighted images of the mice in nasopharyngeal carcinoma xenografted CNE-2 tumors and inferior vena cava. Taylor et al. covalently linked Gd-chelates to the surface of MSNPs via siloxane bonds (Figure 4.7, B) (Taylor et al. 2008). The resulting functionalized MSNPs exhibited exceptionally high relaxivities on both a per-Gd and a per-particle basis. Their efficacy as contrast agents for optical and MR imaging was clearly demonstrated in vitro and in vivo.

Carniato et al. also reported the synthesis of MSNPs as carriers of Gd(III) complexes for the development of MRI contrast agents (Carniato et al. 2009,

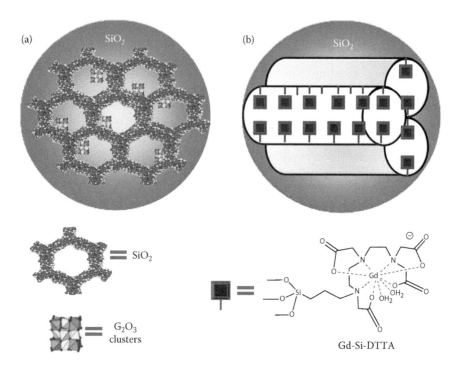

FIGURE 4.7   (See color insert.) Schematic representations of MSNPs designed as highly efficient MRI contrast agents by: (a) incorporating Gd$_2$O$_3$clusters on the inner surface of mesopores (Li, Liu et al. 2011) or (b) functionalizing silica walls with Gd-Si-DTTA complex (Si-DTTA = 3-aminopropyl(trimethoxysilyl) diethylenetriamine tetraacetic acid) via siloxane linkage. *Source*: Taylor et al. (2008).

2010). In a first approach, these authors prepared two novel hybrid materials based on different mesoporous silicas (SBA-15 and nanosized MCM-4l) grafted with DOTA-monoamide Gd complexes (H4DOTA = 1,4,7,10-tetraazacyclododecane-N,N′,N″,N‴-tetraacetic acid) (Carniato et al. 2009). Such work demonstrated that the mesoporous supports played a relevant role in the physicochemical character- istics of the final hybrid materials, since the distribution of the Gd(III) complexes on the inner and/or outer surface was dependent on the pore size and noticeably influenced their relaxometric properties. More specifically, high longitudinal molar relaxivity was estimated for the materials functionalized with Gd(III) com- plexes grafted to the external surface of the particles, due to a high accessibil- ity of the water molecules to the paramagnetic center. Such preliminary studies stimulated further studies on the effects of the chemical nature of the organo- silica surface on the magnetic properties of different Gd(III) chelates immobi- lized on MSNPs (Carniato et al. 2010). For this reason, in addition to the neutral Gd-DOTA monoamide complex ([1] Figure 4.8), two Gd(III) complexes with different structures and magnetic properties were chosen for the functionaliza- tion of the MSNPs. One was the negatively charged Gd-DOTAGA (DOTAGA = 2-(*R*)-2-(4,7,10-triscarboxymethyl-1,4,7,10-tetraazacyclododec-l-yl)pentanedioic

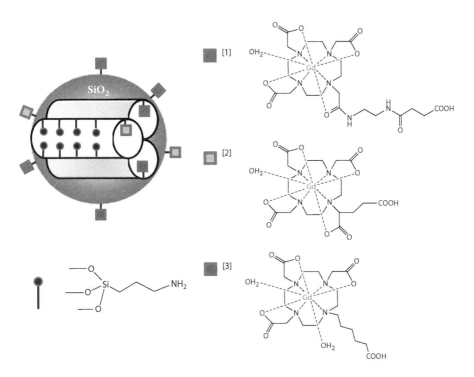

FIGURE 4.8 NH$_2$-functionalized MSNPs immobilizing three Gd(III) complexes based on the 1,4,7,10-tetraazacyclododecane scaffold: Gd-DOTA monoamide [1], Gd-DOTAGA [2], and Gd-DO3A [3]. The immobilization of the three ligands into the MSNPs takes place by reaction between the free carboxylic group of the Gd(III) chelates and the –NH$_2$ functionalities present in the surface of the silica matrix. *Source*: Carniato et al. (2010).

acid) ([2] Figure 4.8). The other one is a neutral Gd-DO3A derivative (H$_3$DO3A = 1,4,7,10-tetraazacyclododecane-1,4,7–triacetic acid) ([3] Figure 4.8). The three ligands were anchored to the MSNPs by reaction between the free carboxylic group of the Gd(III) chelates and the –NH$_2$ functionalities present in surface of silica matrix (Figure 4.8) to form amide groups. Characterization of the resulting nanosystems demonstrated that the Gd(III) complexes were anchored mainly on the external surface of the nanoparticles. However, results of analyses indicated that these ligands would be partially blocking the mesopore entrances. Moreover, the interaction between Gd(III) chelates and surface functional groups markedly influenced the relaxometric properties of the hybrid materials, and was greatly modified passing from ionic NH$_3^+$ to neutral amides. Thus this study demonstrates that the type of Gd(III) chelate and the chemical nature of the support play a key role in determining the magnetic relaxation efficiency of paramagnetically labeled MSNPs.

Recently, Lin and coworkers synthesized a multifunctional MSNP-based MRI contrast agent that can be quickly excreted by the renal pathway after imaging (Vivero-Escoto et al. 2011). The MSNPs contained Gd(III) chelates that were covalently linked via a redox-responsive disulfide moiety. The MSNPs were further functionalized with polyethylene glycol (PEG) and an anisamide ligand to improve their biocompatibility and target specificity. The effectiveness of MSNPs as an MRI imaging contrast agent and their targeting ability were successfully demonstrated in vitro using human colon adenocarcinoma and pancreatic cancer cells. Finally, the capability of this platform as an in vivo MRI contrast agent was tested using a 3T scanner. The Gd(III) chelate was quickly cleaved by the blood pool thiols and eliminated through the renal excretion pathway.

The capacity to monitor cell trafficking and biodistribution in vivo is a prerequisite for developing effective stem cell therapies. Thus MR is an ideal noninvasive imaging technique for tracking stems, although cells must be magnetically labeled by endocytic internalization. Therefore in the search for more efficient noninvasive cellular-internalizing platforms, MSNPs have been proposed as multimodal contrast agents for tracking stem cells (Hsiao et al. 2008; H.-M. Liu et al. 2008). The MSNP-based dual-modal platform combines a fluorescent dye (FITC) and an MR contrast agent. Both T_1- and T_2-weighted MR contrast agents have been incorporated into this system. Gd-based chelates were grafted onto MSNPs to afford Gd-dye MSNPs (Hsiao et al. 2008), and superparamagnetic iron oxide nanoparticles acted as T_2 contrast agents to afford mag-dye MSNPs (H.-M. Liu et al. 2008). Both systems were efficiently internalized into human mesenchymal stem cells (hMSCs) without affecting cell viability, growth, or differentiation. The efficient hMSC tracking was visualized in vitro and in vivo by using a clinical 1.5 T MRI system. In vivo, the labeled cells remained detectable by MRI after long-term growth or differentiation, providing further evidence of the biocompatibility and durability of both of the mag-dye and Gd-dye MSNP nanoprobes. In addition, Mou and coworkers used the dual-modality mag-dye MSNP system to monitor the biodistribution in vivo after eye vein injection in a mouse model (Sudimack and Lee 2000). T_2-weighted MR images revealed that the MSNPs accumulated immediately in the liver, spleen, and kidneys mainly following a vascular mechanism, and signal darkening was observed mainly in the liver and spleen due to nanoparticle accumulation within the reticuloendothelial system at later time points post injection.

4.5 HYPERTHERMIA-BASED THERAPY

Hyperthermia as antitumor therapy consists of heating a tumor region to inhibit the regulatory and growth processes of cancerous cells with the aim of destroying or making them more sensitive to the effects of radiation and antineoplastic drugs (Van Landeghem et al. 2009; Vauthier, Tsapis, and Couvreur 2010). The incorporation of enough maghemite (γ-Fe_2O_3) nanoparticles encapsulated within a mesoporous silica matrix afforded materials that can reach temperatures in the range of hyperthermia under the action of alternating magnetic fields (AMFs) (Figure 4.9) (Ruiz-Hernández et al. 2007, 2008; Arcos et al. 2012). The incorporation of these iron oxide thermoseeds into MSNPs would permit the synergistic combination of hyperthermia and chemotherapy for cancer treatment. Both the mesoporous order and the magnetic properties of the resulting material can be tailored by varying the surfactant/silica molar ratio, the type of surfactant employed, and the number of encapsulated

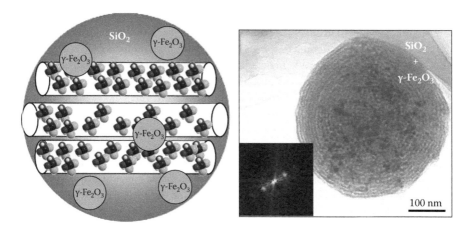

FIGURE 4.9 **(See color insert.)** Schematic representation of MSNPs encapsulating γ-Fe_2O_3 nanoparticles and loaded with drugs. Transmission electron microscopy (TEM) image of the system and its corresponding Fourier transform are also displayed.

magnetic nanoparticles. The capacity of these highly magnetic MSNPs to conduct magnetic hyperthermia upon exposure to a low-frequency AMF has been demonstrated in vitro using human cells of cancerous nature (Martín-Saavedra et al. 2010). Magnetic hyperthermia experiments showed the ability to control the temperature rise in the cell culture environment upon the treatment with magnetic MSNPs and AMF exposure, thus generating heat treatments that severely compromise cell survival. Maximum temperature in MMS suspensions increased to a range above 42°C as a function of the number of particles exposed to AMF. Cell culture experiments evidenced that by adjusting the number of magnetic MSNPs and the time of exposure to AMF, heat treatments of mild to very high intensities could be achieved. Cell viability dropped as a function of the intensity of the heat treatment achieved by magnetic MSNPs and AMF exposures. The possibility of fine tuning the heating power output, together with efficient uptake by tumor cells in vitro, makes magnetic MSNPs promising agents for hyperthermia treatments combined with intracellular delivery of chemotherapeutic drugs.

4.6 BIOSENSING

Detection of biological activity is an attractive diagnostic application. Because of their size and versatile chemistry, nanomaterials are receiving growing attention as powerful tools for biosensing applications (Davis, Chen, and Shin 2008; Ferrari 2005). In comparison with other solid nanoparticle biosensors, MSNPs offer two main advantages that make them highly suitable for biosensing. The first advantage concerns their large surface areas, which allow for the incorporation of a great number of receptors/sensors into the porous matrix and the conjugation with high concentration of analyte molecules to achieve low detection limits. In addition, the fast diffusion of the analytes through the mesopores of MSNPs to the sensor sites also permits high detection rates. The second advantage regards their optical transparency, which permits high detection through layers of the material itself.

Drugs, enzymes, antibodies, and nucleotides, all suitable for MSNP incorporation, have been used as specific molecular receptors to achieve analyte selectivity in biosensing applications (Rosenholm 2011; Slowing et al. 2007; Trewyn et al. 2007). For instance, the simultaneous immobilization of two enzymes, horseradish peroxidase and glucose oxidase, into MSNPs permitted the development of a selective sensor for glucose (Wei et al. 2002). More recently, a hybrid system consisting of MSNPs incorporating poly(amidoamine) dendrimer-encapsulated platinum nanoparticles and immobilizing glucose oxidase was used to modify a glassy carbon electrode for detecting glucose (Han et al. 2011). Another glucose biosensor was developed by immobilizing Pt nanoparticles and glucose oxidase into amino-modified MSNPs (H. Li et al. 2011). Kim et al. (M. Kim et al. 2011) have developed glucose and cholesterol biosensors by developing a nanostructured multicatalyst system consisting of Fe_3O_4 magnetic nanoparticles as peroxidase mimetics and an oxidative enzyme (glucose oxidase or cholesterol oxidase) entrapped in mesoporous silica. Dai et al. reported the immobilization of myoglobin and hemoglobin in mesoporous silica-modified electrodes to be used as a sensor for H_2O_2 and NO_2^- (Dai et al. 2004; Dai, Xu, and Ju 2004). Other researchers have developed alternative approaches to avoid the use of biomolecules to develop biosensors. Thus Radu et al. developed a molecular recognition system capable of distinguishing between structurally similar amino-containing neurotransmitters, such as dopamine and glutamic acid, using different functionalization of the outer and the inner particle surfaces of MSNPs (Radu, Lai, Wiench et al. 2004). The functionalization affected the diffusional penetration rate and thus the detection of analytes into the particles based on their charge. Martínez-Mañez and coworkers prepared MSNPs functionalized with aminomethylanthracene groups to develop another biosensor useful in the recognition and the detection of anions (Descalzo et al. 2002). The bulkiness of the grafted group combined with the steric restrictions provided by the pore size of the material led to the ability of the material to respond in different degrees to ATP, ADP, and AMP (adenosine tri-, di-, and mono-phosphate, respectively).

Current research efforts are being devoted to design-improved DNA biosensors based in MSNPs. Hence, Du et al. developed an integrated sensing system for detection of DNA using new parallel-motif DNA triplex system and graphene-mesoporous silica-gold nanoparticles hybrids (GSGHs) (Du et al. 2012). Such GSGHs offer several advantages for the design of DNA-based electrochemical analytical devices: (1) the Au nanoparticles on the surface of GSGHs can be easily functionalized by SH-DNA; (2) mesoporous silica can reduce the nonspecific adsorption of the DNA; (3) GSGHs may have electronic conductivity that facilitates the electronic transfer of probe molecules; and (4) GSGHs can enlarge the electrode area, which is an important factor for enhancing the detection performance. Considering all these advantages, GSGHs were used as an enhanced element of the integrated sensing platform for the ultrasensitive and -selective detection of DNA through the use of strand-displacement DNA polymerization and parallel-motif DNA triplex system as dual amplifications. In addition such a sensing system could be achieved in the indium tin

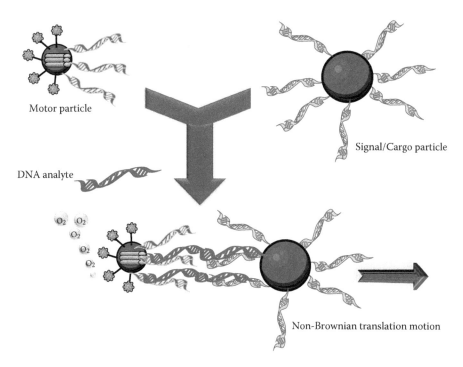

FIGURE 4.10 Schematic representation of motor nanoparticle consisting of MSNPs asymmetrically functionalized by the attachment of single-stranded DNA onto one of its faces, while catalase is immobilized on the other face. The motor particles are able to capture and transport cargo particles functionalized with a noncomplementary single-stranded DNA molecule only if a specific oligonucleotide sequence is present in the media. Catalase enzyme allows catalytic decomposition of hydrogen peroxide to oxygen and water, giving rise to the driving force for the motion of the whole system. *Source*: Simmchen et al. (2012).

oxide (ITO) electrode array, which is of paramount importance for possible multiplex analysis in lab-on-chip.

Vallet-Regí and coworkers have developed a straightforward strategy to design a novel motion-based DNA sensor (Simmchen et al. 2012), The system was based on a mesoporous silica motor nanoparticle particle, which was asymmetrically functionalized by the attachment of single-stranded DNA onto one of its faces, while catalase was immobilized on the other face (Figure 4.10). This enzyme allows catalytic decomposition of hydrogen peroxide to oxygen and water, giving rise to the driving force for the motion of the whole system. On the other hand, the oligonucleotide strand permits the selective capture of DNA-functionalized particles (cargo). The two DNA strands are not complementary to each other, but each is complementary to one of the terminal fragments of a larger strand (analyte) that acts as bridge between the particles, because of the high selectivity of the DNA hybridization process. Without hybridization on both sides of the analyte DNA, the motor is not connected to the detection system (cargo) and no easily detectable movement is detected by optical microscopy because the larger cargo particles do not move (Figure 4.10). In this way, the novel self-propelled nanodevice not only performs motion and cargo transport but also allows us to accomplish motion-based detection of oligonucleotide sequences by direct visual tracking of the cargo. This new strategy appears very promising for the design of lab-on-chip devices for pathogen DNA detection or other interesting applications (Simmchen et al. 2012).

4.7   THERANOSTIC NANOMEDICINE

In recent years nanomedicine has evolved to encompass integrated, biocompatible multifunctional nanoparticle platforms that play dual roles as diagnostic and therapeutic agents (Lammers et al. 2011; Choi et al. 2012). The field has experienced such mushrooming expansion that the term *theranostic*, derived from thera(py) + (diag)nostics, has been coined to link the fields of diagnostics and therapeutics. Multifunctional nanoparticles that combine imaging and therapeutic agents triggering gene or drug release at target sites when exposed to external stimuli can be very powerful for understanding and visualizing in real-time drug delivery, drug release, and drug efficacy. Although more research is needed before theranostic nanomedicine can be implemented in the clinic, current research indicates that theranostic nanomedicine may revolutionize the diagnosis and treatment of many diseases and help to realize the potential of personalized medicine.

In the search for novel and efficient therapies for the treatment of several diseases, including cancer, photodynamic therapy (PDT) has emerged as an alternative to chemotherapy and radiotherapy (Juarranz et al. 2008; Robertson, Evans, and Abrahamse 2009). It involves the use of light and photosensitizers (PSs) that accumulate in the tumor tissue (Figure 4.11). PSs are drugs that can transfer their energy from their triplet excited state to neighboring oxygen molecules when activated by light of a specific wavelength (Ortel, Shea, and Calzavara-Pinton 2009). Single oxygen ($^1O_2$) and other cytotoxic reactive oxygen species (ROSs) are formed and lead to the destruction of cancer cells by both apoptosis and necrosis (Berg et al. 2005). However, the role of PSs is not restricted exclusively to the therapeutic generation of singlet oxygen. Many PSs also act as bright fluorophores, and some of them emit in the NIR region of the spectra, which is preferred for in vivo OI. Thus nanoparticles carrying photosensitizers can be used as theranostic agents for PDT.

MSNPs are particularly suitable for the development of theranostic nanoparticles. Their unique features allow the design of multifunctional platforms for the incorporation of a wide range of functional molecules, such as contrast agents and therapeutic and biomolecular targeting ligands (Vivero-Escoto, Huxford-Phillips, and Lin 2012; Lee, Lee, Kim et al. 2011; Couleaud et al. 2010).

Cheng et al. developed tri-functionalized MSNPs for use as theranostic agents that orchestrate the trio of imaging, targeting, and therapy in a single particle (S.-H. Cheng et al. 2010). An NIR dye (ATTO647N) was directly incorporated into the silica network during the MSNPs' synthesis. Then the photosensitizer *meso*-tetratolylporphyrin-Pd (PdTPP) was grafted within the mesopores of MSNPs to enable PDT. Finally, the exterior surface of the MSNPs was functionalized with cyclic RGD peptides attached to a PEG chain and was used for targeting the overexpressed $\alpha_v\beta_3$ integrins of cancer cells and to ensure the internalization of the photosensitizer PdTPP (Figure 4.12) (S.-H. Cheng et al. 2010). In vitro cell evaluation of

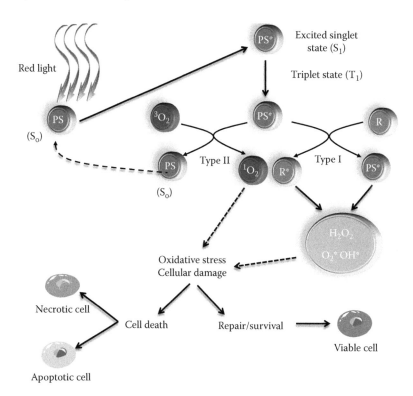

FIGURE 4.11 PDT requires three main elements: a photosensibilizer (PS), light, and oxygen. When exposed to specific wavelengths of light, the PS in its singlet ground state (S_0) becomes activated to an excited singlet state (S_1), which is followed by intersystem crossing to an excited triplet state (T_1). Transfer energy from T_1 to biological substrates and molecular oxygen, via type I and II reactions, generates ROS ($^1O_2$, H_2O_2, $O_2^{\bullet-}$, $\bullet OH$). They produce cellular damage, which can be repaired or can kill tumor cells mainly by necrosis or apoptosis. *Source*: Juarranz et al. (2008).

the theranostic platform demonstrated not only excellent targeting specificity and minimal collateral damage but highly potent therapeutic effect as well.

Brinker and coworkers have developed MSNP-supported lipid bilayers (protocells), which synergistically combine properties of liposomes and MSNP particles as theranostic platforms (Ashley et al. 2011, 2012; Liu et al. 2009). Protocells modified with a targeting peptide that binds to human hepatocellular carcinoma exhibit a 10,000-fold greater affinity for human hepatocellular carcinoma than for hepatocytes, endothelial cells, or immune cells. Furthermore, protocells were loaded with combinations of therapeutic (drugs, small interfering RNA, and toxins) and diagnostic (quantum dots) agents and modified to promote endosomal escape and nuclear accumulation of selected cargos. The enormous capacity of the high surface area of the mesoporous core combined with the enhanced targeting efficacy enabled by the fluid-supported lipid bilayer enabled a single protocell loaded with a drug cocktail to kill a drug-resistant human hepatocellular carcinoma cell, representing a 10^6-fold improvement over comparable liposomes.

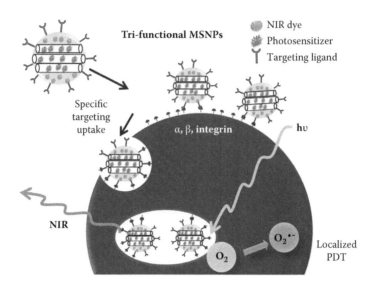

FIGURE 4.12 (**See color insert.**) Schematic representation of tri-functional MSNPs as theranostic agents. The presence of RGD peptides grafted to the external surface of MSNPs permits targeting the overexpressed $\alpha_v\beta_3$ integrins of cancer cells. The NIR dye (ATTO647N) incorporated within the silica network allows traceable imaging of particles. The photosensitizer *meso*-tetratolylporphyrin-Pd (PdTPP) grafted to the mesopores enables local PDT. *Source*: Cheng et al. (2010).

Recently, Hyeon and coworkers synthesized a monodisperse core-shell platform consisting of a single magnetite (Fe_3O_4) nanocrystal core surrounded by a shell of mesoporous silica (Fe_3O_4 MSNPs) (Figure 4.13) (Kim et al. 2008). This system was then functionalized with optical agents (FITC and rhodamine B) and PEG chains.

The OI capability of this material was used to determine its internalization in cancer cells. In addition, the MRI T_2-weighted properties of Fe_3O_4 MSNPs-PEG were measured with a 1.5 T scanner. The fluorescent and T_2-weighted MR images of phantoms showed that as the concentration of the nanoparticles was increased, a brighter fluorescence of rhodamine B and a darker T_2 signal were observed, proving that the uptake of the particles was concentration dependent. To examine the performance of Fe_3O_4 MSNPs-PEG as a drug delivery vehicle, doxorubicin (DOX), a potent anticancer agent, was loaded and the cytotoxic effect of this system was tested on SK-BR-3 cells. The cytotoxic efficacy of the DOX-loaded Fe_3O_4 MSNPs-PEG increased as the concentration of DOX-loaded nanoparticles was increased, whereas Fe_3O_4 MSNPs-PEG alone did not show cytotoxicity to cancer cells even at high concentration. This result indicates that the Fe_3O_4 MSNPs system has a potential for drug loading and delivery into cancer cells to induce cell death. The authors investigated the in vivo imaging capabilities of this nanosystem in a breast cancer xenograft model. After 2 hours post-injection, the accumulation of nanoparticles in the tumor was detected by T_2-weighted MR. The accumulation of Fe_3O_4 MSNPs-PEG was further confirmed by fluorescence imaging of the tumor and organs of mice sacrificed 24 hours after injection.

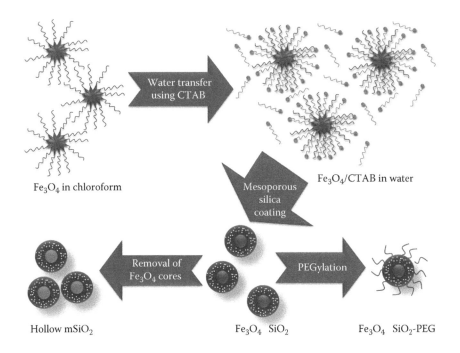

FIGURE 4.13   Schematic illustration of the synthetic procedure for the magnetite nanocrystal/mesoporous silica core-shell nanoparticles. *Source*: Kim et al. (2008).

The same authors also reported the synthesis of uniform dye-doped MSNPs immobilized with multiple magnetite nanocrystals on the surface (designated as Fe_3O_4 MSNPs) as well as their applications to simultaneous MRI, to fluorescence imaging, and as a drug delivery vehicle (Lee, Lee et al. 2009) (Figure 4.14).

The superparamagnetic properties of the Fe_3O_4 nanocrystals enabled the system to be used as a contrast agent in MRI, and the fluorophore (FITC) into the silica framework imparted optical imaging modality. Integrating a multitude of Fe_3O_4 nanocrystals on the silica surface resulted in remarkable enhancement of MR signal due to the synergistic magnetism. The anticancer drug DOX was loaded into the mesopores (Figure 4.14) and the antitumor efficacy of this vehicle was successfully tested in vitro. In vivo passive targeting and accumulation of the nanoparticles at the tumor sites was confirmed by both T_2 MR and fluorescence imaging. Furthermore, apoptotic morphology was clearly detected in tumor tissues of mice treated with DOX-loaded nanocomposite nanoparticles, demonstrating that DOX was successfully delivered to the tumor sites and its anticancer activity was preserved. These authors also developed multifunctional nanoparticles for simultaneous fluorescence and MR imaging, and pH-sensitive drug release (Lee, Lee, Lee et al. 2011). This system was fabricated by immobilizing pH-responsive hydrazone bonds, magnetite nanoparticles, and fluorescent dyes in mesoporous silica nanoparticles. pH-dependent release of doxorubicin was demonstrated, and the nanocomposite nanoparticles showed fluorescence emission and MR contrast effects.

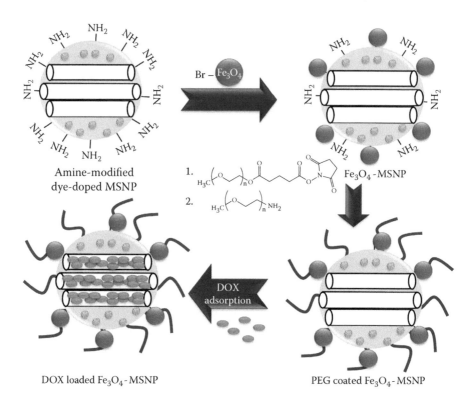

FIGURE 4.14 Schematic representation of the synthesis of Fe_3O_4-MSNPs functionalized with PEG chains and loaded with DOX for simultaneous MRI, fluorescence imaging, and a drug delivery. *Source*: Lee, Lee et al. (2009).

The same research group has recently developed multifunctional core-shell-structured MSNPs for simultaneous MR and fluorescence imaging, cell targeting, and photodynamic therapy (PDT) as a promising material for cancer diagnosis and therapy (F. Wang, Chen et al. 2011). The encapsulation of a single Fe_3O_4 magnetite nanoparticle and fluorescence dyes (FITC) in one nonporous silica core endowed the nanoparticles with the MRI and fluorescence imaging capabilities, allowing noninvasive tracking and monitoring of the nanoparticles within cells and even the body. The photosensitizer molecules (AlC4Pc), which were covalently bound to the mesoporous silica shell, exhibited a good stability against leaching and an excellent efficiency in photogeneration of reactive oxygen species. Furthermore, the surface modification of the core-shell nanoparticles by folic acid allowed the targeted delivery of the AlC4Pc to the cancer cells and therefore minimized the toxicity to the surrounding healthy tissues.

Recently, Wang et al. (T. Wang, Zhang et al. 2011) have developed highly uniform and multifunctional hollow MSNPs with cell-imaging capability for dual-mode cancer treatment using PDT and chemotherapy. MSNPs comprised a fluorescence imaging agent (FITC) for optical tracking and hematoporphyrin as photosensitizer for PDT, both of them covalently linked to the silica walls of MSNPs. Then

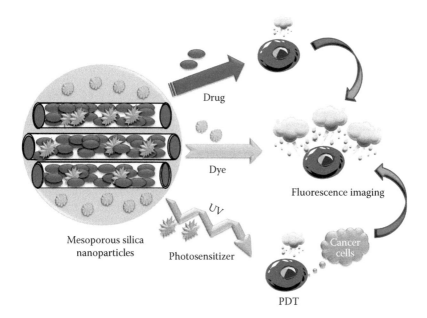

Drug

Dye

Fluorescence imaging

UV

Mesoporous silica nanoparticles

Photosensitizer

Cancer cells

PDT

FIGURE 4.15 (**See color insert.**) MSNPs for theranostic applications, combining fluorescence imaging (by incorporation of a fluorescent dye) and dual-mode cancer therapy (chemotherapy and photodynamic therapy [PDT]).The synergy of combining chemotherapy and PDT exhibiting high therapeutic efficacy for cancer cells is also schematically displayed.

the chemotherapeutic drug DOX was loaded into the mesoporous cavities for drug delivery. These multifunctional MSNPs exhibited good biocompatibility for, and cell uptake by, cancer cells, which can guarantee the therapeutic efficacy as well as facilitate the application of traceable imaging. In vitro cell assays to evaluate the cancer therapy of the DOX-loaded MSNPs demonstrated that a cooperative therapeutic system combining chemotherapy and PDT produced remarkable therapeutic efficacy relative to the individual means (Figure 4.15). Further in vivo studies are under way to optimize all the parameters that provide the best synergistic effect.

Tamanoi and coworkers reported the synthesis of MSNP-based theranostic particles containing an OI agent (FITC) and an antitumor drug (camptothecin, CPT) (J. Lu et al. 2010). The authors evaluated the toxicity, biodistribution, clearance, and therapeutic properties of MSNPs. They evaluated the short- and long-term toxicities of these nanosystems. The biodistribution and excretion of MSNPs were determined by OI and inductively coupled plasma optical emission spectrometry (ICP-OES), respectively. Fluorescent images showed that MSNPs mainly accumulate in the tumor, kidneys, and liver. Both renal and hepatobiliary routes of excretion were investigated. The results derived from the urine and feces showed that nanoparticles are cleared through both pathways. To assess the therapeutic efficacy of CPT-loaded MSNPs in vivo, a xenograft model of MCF-7 human breast cancer cells was used. Results revealed significant tumor growth inhibition with the theranostic particles at the end of the experiment.

4.9   BIOCOMPATIBILITY AND TOXICITY

The application of MSNPs in nanomedicine requires previous research efforts to evaluate their cellular uptake and biocompatibility in vitro, and toxicity, biodistribution, and clearance in vivo.

The cellular uptake of MSNPs can be investigated by grafting fluorescent dye molecules to the MSNPs, which enable visualizing the nanosystems by fluorescence and confocal microscopy, as commented on in Section 4.4.1 (Y.-S. Lin et al. 2005, 2006; Liong et al. 2009; J. Lu, Liong, Sherman et al. 2007; J. Lu, Liong, Zink et al. 2007; Slowing, Trewyn, and Lin 2006). The efficient cellular uptake of MSNPs and their good biocompatibility have been established by several research groups using both healthy and cancer cell lines (J. Lu, Liong, Sherman et al. 2007; Radu, Lai, Jeftinija et al. 2004; Vivero-Escoto et al. 2010; Hudson et al. 2008). No cellular toxicity is observed up to 100 µg.mL for unmodified MSNPs with 100 nm of particle size (Hudson et al. 2008; J. Lu et al. 2008, 2010; Meng et al. 2010; Thomas et al. 2010), which is above the effective particle concentration needed for most therapeutic treatments. The cellular uptake and cytotoxicity of MSNPs depend on the particle's size, shape, surface charge, and functional groups, as reported by different scientific groups (Vivero-Escoto et al. 2010; Wu, Hung, and Mou 2011).

Cell staining assays together with the fluorescently labeled MSNPs permit the study of the cellular uptake mechanisms and the particle location inside the cells. Small MSNPs (<200–300 nm) are generally taken up by endocytosis. Details of the different endocytosis mechanisms for MSNPs can be found in recent reviews (Vivero-Escoto et al. 2010; Wu, Hung, and Mou 2011). Briefly, the internalization of MSNPs through endocytosis results in the formation of vesicles that capture the nanoparticles in the extracellular environment. Then these vesicles undergo a complex series of fusion events directing the internalized MSNPs to the cytosolic compartment. The succession of events after the MSNPs have been endocytosed can be divided into the following sequence. The material is first transported to primary endosomes followed by transport to sorting endosomes. From sorting endosomes, a fraction of the MSNPs is directed back to the cell exterior through recycling endosomes, while the remaining fraction is transported to secondary endosomes, which fuse with the acidic lysosomes. The MSNPs end up inside these acidic (pH ≈ 4.5) organelles in cells. It has also been demonstrated that nanoparticles with surface groups that can be protonated assist the "proton sponge effect," which leads to the endosomal escape of the uptaken particles (Xia et al. 2009). This permits the membrane-impermeable molecules such as hydrophilic drugs, DNA, and siRNA to be released from the membrane-bound endosomes and travel to their effective sites.

Several studies of the in vivo toxicity, biodistribution, and clearance of MSNPs in animal models have been reported (J. Lu et al. 2008, 2010; X. Huang, Li et al. 2011; Lin, Hurley, and Haynes 2012). Further, 100 nm MSNPs of hollow and MCM-41 types have shown very good long-term (>1 month) biocompatibility in mice (J. Lu et al. 2010). The in vivo toxicity of MSNPs depends on the injection method. The particle size and surface functional groups and charges greatly affect the biodistribution and pharmacokinetics (Souris et al. 2010; Van Schooneveld et al. 2010; He and

Shi 2011). For instance, positively charged MSNPs have a much faster clearance rate than negatively charged ones. This fact was demonstrated by Souris and coworkers (C.-H. Lee, Cheng et al. 2009; Souris et al. 2010), who investigated the factors that govern the rethiculoendothelial system (RES) escape and subsequent hepatobiliary excretion of MSNPs higher than 80 nm. NIR fluorescent dye (ICG) was selected to label MSNPs for use as in vivo contrast agents. Two different routes, covalent and ionic bonding, were followed to incorporate IGF into MSNPs, which resulted in comparable systems of significantly different net surface change. Thus ICG was covalently linked to aminopropyl-functionalized MSNPs (MSNPs-NH2-ICG) or ionically conjugated to trimethylamonium-functionalized MSNPs (MSNPs-TA-ICG). Zeta potential measurements of both systems indicated that at the physiological pH of 7.4, $MSNPs-NH_2-ICG$ possess more surface charge (+34.4 mV) than MSNPs-TA-ICG (–17.4 mV). In vivo fluorescence imaging of mice and ex vivo fluorescence imaging of their harvested tissues and excrement revealed remarkable differences in bioelimination of both systems. Results indicated that $MSNPs-NH_2-ICG$ underwent rapid transport into the gastrointestinal tract and subsequent elimination via feces, showing rapid onset of clearance within 30 minutes after administration. On the contrary, MSNPs-TA-ICG exhibited much longer residence in the body for days to months, possessing high uptake and retention in the liver. Analysis of organs harvested three days after injection suggested a possible oncoming of MSNP biodegradation into orthosilicic acid with likely renal excretion. Results indicated that both $MSNPs-NH_2-ICG$ and MSNPs-TA-ICG systems experienced sequestration by the liver. However, the authors postulate that the greater surface charge of the former would be favoring its higher opsonization by serum proteins and thus its hepatobiliary excretion into the gastrointestinal tract. Therefore the main findings derived from such investigation suggest that charge-dependent adsorption of serum proteins considerably facilitates the hepatobiliary excretion of silica nanoparticles, and that the nanoparticle residence time in vivo can be regulated by controlling the surface characteristics. As described in Section 4.2.1, it is possible to coat MSNPs with PEG to increase their in vivo circulation time by preventing their removal by phagocytes. Accordingly, fewer PEG-coated MSNPs are trapped in the RES of the liver and spleen. The in vivo degradation and urinary excretion of MSNPs have also been observed in several cases. As commented on in Section 4.2.2, a paramount issue to consider when designing in vivo drug delivery platforms is the targeting of diseased organs or tissues. Thus targeting of MSNPs with moieties that selectively bind cell surface receptors to trigger receptor-mediated endocytosis increases the cellular uptake of the nanosystems (Rosenholm et al. 2008; Slowing, Trewyn, and Lin 2006; Tsai et al. 2009; Ferris et al. 2011). The in vivo capability of delivering anticancer drugs, such as CPT or DOX, for tumor shrinking has been also demonstrated into human xenografts in mice (J. Lu et al. 2010; Meng et al. 2011).

REFERENCES

Alexis, Frank, Eric Pridgen, Linda K. Molnar, and Omid C. Farokhzad. 2008. "Factors Affecting the Clearance and Biodistribution of Polymeric Nanoparticles." *Molecular Pharmaceutics* 5 (4):505–15.

Ambrogio, Michael W., Courtney R. Thomas, Yan-Li Zhao, Jeffrey I. Zink, and J. Fraser Stoddart. 2011. "Mechanized Silica Nanoparticles: A New Frontier in Theranostic Nanomedicine." *Accounts of Chemical Research* 44 (10):903–13.

Andersson, Jenny, Jessica Rosenholm, Sami Areva, and Mika Lindén. 2004. "Influences of Material Characteristics on Ibuprofen Drug Loading and Release Profiles from Ordered Micro- and Mesoporous Silica Matrices." *Chemistry of Materials* 16 (21):4160–67.

Anglin, Emily J., Lingyun Cheng, William R. Freeman, and Michael J. Sailor. 2008. "Porous Silicon in Drug Delivery Devices and Materials." *Advanced Drug Delivery Reviews* 60 (11):1266–77.

Arcos, D., A. López-Noriega, E. Ruiz-Hernández, O. Terasaki, and M. Vallet-Regí. 2009. "Ordered Mesoporous Microspheres for Bone Grafting and Drug Delivery." *Chemistry of Materials* 21 (6):1000–1009.

Arcos, Daniel, Vanesa Fal-Miyar, Eduardo Ruiz-Hernandez, Mar Garcia-Hernández, M. Luisa Ruiz-Gonzalez, Jose Gonzalez-Calbet, and María Vallet-Regí. 2012. "Supramolecular Mechanisms in the Synthesis of Mesoporous Magnetic Nanospheres for Hyperthermia." *Journal of Materials Chemistry* 22 (1):64–72.

Ashley, Carlee E., Eric C. Carnes, Katharine E. Epler, David P. Padilla, Genevieve K. Phillips, Robert E. Castillo, Dan C. Wilkinson, Brian S. Wilkinson, Cameron A. Burgard, Robin M. Kalinich, Jason L. Townson, Bryce Chackerian, Cheryl L. Willman, David S. Peabody, Walker Wharton, and C. Jeffrey Brinker. 2012. "Delivery of Small Interfering RNA by Peptide-Targeted Mesoporous Silica Nanoparticle-Supported Lipid Bilayers." *ACS Nano* 6 (3):2174–88.

Ashley, Carlee E., Eric C. Carnes, Genevieve K. Phillips, David Padilla, Paul N. Durfee, Page A. Brown, Tracey N. Hanna, Juewen Liu, Brandy Phillips, Mark B. Carter, Nick J. Carroll, Xingmao Jiang, Darren R. Dunphy, Cheryl L. Willman, Dimiter N. Petsev, Deborah G. Evans, Atul N. Parikh, Bryce Chackerian, Walker Wharton, David S. Peabody, and C. Jeffrey Brinker. 2011. "The Targeted Delivery of Multicomponent Cargos to Cancer Cells by Nanoporous Particle-Supported Lipid Bilayers." *Natural Materials* 10 (5):389–97.

Avnir, David, Thibaud Coradin, Ovadia Lev, and Jacques Livage. 2006. "Recent Bio-applications of Sol-Gel Materials." *Journal of Materials Chemistry* 16 (11):1013–30.

Barbé, C., J. Bartlett, L. Kong, K. Finnie, H. Q Lin, M. Larkin, S. Calleja, A. Bush, and G. Calleja. 2004. "Silica Particles: A Novel Drug-Delivery System." *Advanced Materials* 16 (21):1959–66.

Berg, K., P. K. Selbo, A. Weyergang, A. Dietze, L. Prasmickaite, A. Bonsted, B. Ø Engesaeter, E. Angell-Petersen, T. Warloe, N. Frandsen, and A. HØGset. 2005. "Porphyrin-Related Photosensitizers for Cancer Imaging and Therapeutic Applications." *Journal of Microscopy* 218 (2):133–47.

Bhattarai, Shanta, Elayaraja Muthuswamy, Amit Wani, Michal Brichacek, Antonio Castañeda, Stephanie Brock, and David Oupicky. 2010. "Enhanced Gene and siRNA Delivery by Polycation-Modified Mesoporous Silica Nanoparticles Loaded with Chloroquine." *Pharmaceutical Research* 27 (12):2556–68.

Boissiere, Cedric, David Grosso, Alexandra Chaumonnot, Lionel Nicole, and Clement Sanchez. 2011. "Aerosol Route to Functional Nanostructured Inorganic and Hybrid Porous Materials." *Advanced Materials* 23 (5):599–623.

Boussif, O., F. Lezoualc'h, M. A. Zanta, M. D. Mergny, D. Scherman, B. Demeneix, and J. P. Behr. 1995. "A Versatile Vector for Gene and Oligonucleotide Transfer into Cells in Culture and In Vivo: Polyethylenimine." *Proceedings of the National Academy of Sciences* 92 (16):7297–7301.

Brevet, David, Magali Gary-Bobo, Laurence Raehm, Sebastien Richeter, Ouahiba Hocine, Kassem Amro, Bernard Loock, Pierre Couleaud, Celine Frochot, Alain Morere, Philippe Maillard, Marcel Garcia, and Jean-Olivier Durand. 2009. "Mannose-Targeted Mesoporous Silica Nanoparticles for Photodynamic Therapy." *Chemical Communications* (12):1475–77.

Brinker, C. Jeffrey, Yunfeng Lu, Alan Sellinger, and Hongyou Fan. 1999. "Evaporation-Induced Self-Assembly: Nanostructures Made Easy." *Advanced Materials* 11 (7):579–85.

Bruhwiler, Dominik. 2010. "Postsynthetic Functionalization of Mesoporous Silica." *Nanoscale* 2 (6):887–92.

Burleigh, Mark C., Michael A. Markowitz, Mark S. Spector, and Bruce P. Gaber. 2001. "Amine-Functionalized Periodic Mesoporous Organosilicas." *Chemistry of Materials* 13 (12):4760–66.

Carniato, Fabio, Lorenzo Tei, Maurizio Cossi, Leonardo Marchese, and Mauro Botta. 2010. "A Chemical Strategy for the Relaxivity Enhancement of GdIII Chelates Anchored on Mesoporous Silica Nanoparticles." *Chemistry: A European Journal* 16 (35):10727–34.

Carniato, Fabio, Lorenzo Tei, Walter Dastru, Leonardo Marchese, and Mauro Botta. 2009. "Relaxivity Modulation in Gd-Functionalised Mesoporous Silicas." *Chemical Communications* (10):1246–48.

Chen, Alex M., Min Zhang, Dongguang Wei, Dirk Stueber, Oleh Taratula, Tamara Minko, and Huixin He. 2009. "Co-delivery of Doxorubicin and Bcl-2 siRNA by Mesoporous Silica Nanoparticles Enhances the Efficacy of Chemotherapy in Multidrug-Resistant Cancer Cells." *Small* 5 (23):2673–77.

Cheng, Shih-Hsun, Chia-Hung Lee, Meng-Chi Chen, Jeffrey S. Souris, Fan-Gang Tseng, Chung-Shi Yang, Chung-Yuan Mou, Chin-Tu Chen, and Leu-Wei Lo. 2010. "Tri-Functionalization of Mesoporous Silica Nanoparticles for Comprehensive Cancer Theranostics—the Trio of Imaging, Targeting and Therapy." *Journal of Materials Chemistry* 20 (29):6149–57.

Cho, Eun Chul, Charles Glaus, Jingyi Chen, Michael J. Welch, and Younan Xia. 2010. "Inorganic Nanoparticle-Based Contrast Agents for Molecular Imaging." *Trends in Molecular Medicine* 16 (12):561–73.

Cho, Kwangjae, Xu Wang, Shuming Nie, Zhuo Chen, and Dong M. Shin. 2008. "Therapeutic Nanoparticles for Drug Delivery in Cancer." *Clinical Cancer Research* 14 (5):1310–16.

Choi, H. S., and J. V. Frangioni. 2010. "Nanoparticles for Biomedical Imaging: Fundamentals of Clinical Translation." *Molecular Imaging* 9 (6):291–310.

Choi, Ki Young, Gang Liu, Seulki Lee, and Xiaoyuan Chen. 2012. "Theranostic Nanoplatforms for Simultaneous Cancer Imaging and Therapy: Current Approaches and Future Perspectives." *Nanoscale* 4 (2):330–42.

Claesson, Per M., Eva Blomberg, Johan C. Fröberg, Tommy Nylander, and Thomas Arnebrant. 1995. "Protein Interactions at Solid Surfaces." *Advances in Colloid and Interface Science* 57:161–227.

Colilla, Montserrat, Miguel Manzano, Isabel Izquierdo-Barba, María Vallet-Regí, Cédric Boissiére, and Clément Sanchez. 2010. "Advanced Drug Delivery Vectors with Tailored Surface Properties Made of Mesoporous Binary Oxides Submicronic Spheres." *Chemistry of Materials* 22 (5):1821–30.

Coti, Karla K., Matthew E. Belowich, Monty Liong, Michael W. Ambrogio, Yuen A. Lau, Hussam A. Khatib, Jeffrey I. Zink, Niveen M. Khashab, and J. Fraser Stoddart. 2009. "Mechanised Nanoparticles for Drug Delivery." *Nanoscale* 1 (1):16–39.

Couleaud, Pierre, Vincent Morosini, Celine Frochot, Sebastien Richeter, Laurence Raehm, and Jean-Olivier Durand. 2010. "Silica-Based Nanoparticles for Photodynamic Therapy Applications." *Nanoscale* 2 (7):1083–95.

Dai, Zhihui, Songqin Liu, Huangxian Ju, and Hongyuan Chen. 2004. "Direct Electron Transfer and Enzymatic Activity of Hemoglobin in a Hexagonal Mesoporous Silica Matrix." *Biosensors and Bioelectronics* 19 (8):861–67.

Dai, Zhihui, Xiaoxing Xu, and Huangxian Ju. 2004. "Direct Electrochemistry and Electrocatalysis of Myoglobin Immobilized on a Hexagonal Mesoporous Silica Matrix." *Analytical Biochemistry* 332 (1):23–31.

Danhier, Fabienne, Olivier Feron, and Véronique Préat. 2010. "To Exploit the Tumor Microenvironment: Passive and Active Tumor Targeting of Nanocarriers for Anti-cancer Drug Delivery." *Journal of Controlled Release* 148 (2):135–46.

Davis, Mark E., Zhuo Chen, and Dong M. Shin. 2008. "Nanoparticle Therapeutics: An Emerging Treatment Modality for Cancer." *Nature Reviews Drug Discovery* 7 (9):771–82.

De, Mrinmoy, Partha S. Ghosh, and Vincent M. Rotello. 2008. "Applications of Nanoparticles in Biology." *Advanced Materials* 20 (22):4225–41.

Dennig, Jörg, and Emma Duncan. 2002. "Gene Transfer into Eukaryotic Cells Using Activated Polyamidoamine Dendrimers." *Reviews in Molecular Biotechnology* 90 (3–4):339–47.

Descalzo, A. B., D. Jimenez, M. D. Marcos, R. Martínez-Máñez, J. Soto, J. El Haskouri, C. Guillém, D. Beltrán, P. Amorós, and M. V. Borrachero. 2002. "A New Approach to Chemosensors for Anions Using MCM-41 Grafted with Amino Groups." *Advanced Materials* 14 (13–14):966–69.

Dosio, Franco, L. Harivardhan Reddy, Annalisa Ferrero, Barbara Stella, Luigi Cattel, and Patrick Couvreur. 2010. "Novel Nanoassemblies Composed of Squalenoyl-Paclitaxel Derivatives: Synthesis, Characterization, and Biological Evaluation." *Bioconjugate Chemistry* 21 (7):1349–61.

Du, Yan, Shaojun Guo, Haixia Qin, Shaojun Dong, and Erkang Wang. 2012. "Target-Induced Conjunction of Split Aptamer as New Chiral Selector for Oligopeptide on Graphene-Mesoporous Silica-Gold Nanoparticle Hybrids Modified Sensing Platform." *Chemical Communications* 48 (6):799–801.

Duan, Hongwei, and Shuming Nie. 2007. "Cell-Penetrating Quantum Dots Based on Multivalent and Endosome-Disrupting Surface Coatings." *Journal of the American Chemical Society* 129 (11):3333–38.

Dubikovskaya, Elena A., Steve H. Thorne, Thomas H. Pillow, Christopher H. Contag, and Paul A. Wender. 2008. "Overcoming Multidrug Resistance of Small-Molecule Therapeutics through Conjugation with Releasable Octaarginine Transporters." *Proceedings of the National Academy of Sciences* 105 (34):12128–33.

Dufort, Sandrine, Lucie Sancey, and Jean-Luc Coll. 2012. "Physico-Chemical Parameters That Govern Nanoparticles Fate Also Dictate Rules for Their Molecular Evolution." *Advanced Drug Delivery Reviews* 64 (2):179–89.

Edinger, Daniel, and Ernst Wagner. 2011. "Bioresponsive Polymers for the Delivery of Therapeutic Nucleic Acids." *Wiley Interdisciplinary Reviews: Nanomedicine and Nanobiotechnology* 3 (1):33–46.

Elnakat, Hala, and Manohar Ratnam. 2004. "Distribution, Functionality and Gene Regulation of Folate Receptor Isoforms: Implications in Targeted Therapy." *Advanced Drug Delivery Reviews* 56 (8):1067–84.

Erbacher, Patrick, Thierry Bettinger, Pascale Belguise-Valladier, Shaomin Zou, Jean-Luc Coll, Jean-Paul Behr, and Jean-Serge Remy. 1999. "Transfection and Physical Properties of Various Saccharide, Poly(ethylene glycol), and Antibody-Derivatized Polyethylenimines (PEI)." *The Journal of Gene Medicine* 1 (3):210–22.

Esfand, R., and D. A. Tomalia. 2001. "Poly(amidoamine) (PAMAM) Dendrimers: From Biomimicry to Drug Delivery and Biomedical Applications." *Drug Discovery Today* 6 (8):427–36.

Esmaeili, Farnaz, Mohammad Hossein Ghahremani, Behnaz Esmaeili, Mohammad Reza Khoshayand, Fatemeh Atyabi, and Rassoul Dinarvand. 2008. "PLGA Nanoparticles of Different Surface Properties: Preparation and Evaluation of Their Body Distribution." *International Journal of Pharmaceutics* 349 (1–2):249–55.

Farokhzad, Omid C., and Robert Langer. 2009. "Impact of Nanotechnology on Drug Delivery." *ACS Nano* 3 (1):16–20.

Faure, Anne-Charlotte, Sandrine Dufort, Véronique Josserand, Pascal Perriat, Jean-Luc Coll, Stéphane Roux, and Olivier Tillement. 2009. "Control of the In Vivo Biodistribution of Hybrid Nanoparticles with Different Poly(ethylene glycol) Coatings." *Small* 5 (22):2565–75.

Ferrari, Mauro. 2005. "Cancer Nanotechnology: Opportunities and Challenges." *Nature Reviews Cancer* 5 (3):161–71.

Ferris, Daniel P., Jie Lu, Chris Gothard, Rolando Yanes, Courtney R. Thomas, John-Carl Olsen, J. Fraser Stoddart, Fuyuhiko Tamanoi, and Jeffrey I. Zink. 2011. "Synthesis of Biomolecule-Modified Mesoporous Silica Nanoparticles for Targeted Hydrophobic Drug Delivery to Cancer Cells." *Small* 7 (13):1816–26.

Feuerbach, Frederick J., and Ronald G. Crystal. 1996. "Progress in Human Gene Therapy." *Kidney International* 49 (6):1791–94.

Florea, Bogdan, Clare Meaney, Hans Junginger, and Gerrit Borchard. 2002. "Transfection Efficiency and Toxicity of Polyethylenimine in Differentiated Calu-3 and Nondifferentiated COS-1 Cell Cultures." *The AAPS Journal* 4 (3):1–11.

Gergely, SzakÁCs, Jakab Katalin, Antal Ferenc, and Sarkadi Balázs. 1998. "Diagnostics of Multidrug Resistance in Cancer." *Pathology & Oncology Research* 4 (4):251–57.

Ghosh, Partha, Gang Han, Mrinmoy De, Chae Kyu Kim, and Vincent M. Rotello. 2008. "Gold Nanoparticles in Delivery Applications." *Advanced Drug Delivery Reviews* 60 (11):1307–15.

Godbey, W. T., Kenneth K. Wu, and Antonios G. Mikos. 1999. "Tracking the Intracellular Path of Poly(ethylenimine)/DNA Complexes for Gene Delivery." *Proceedings of the National Academy of Sciences* 96 (9):5177–81.

González, Blanca, Montserrat Colilla, Carlos López de Laorden, and María Vallet-Regí. 2009. "A Novel Synthetic Strategy for Covalently Bonding Dendrimers to Ordered Mesoporous Silica: Potential Drug Delivery Applications." *Journal of Materials Chemistry* 19 (47):9012–24.

González, Blanca, Eduardo Ruiz-Hernández, María José Feito, Carlos Lopez de Laorden, Daniel Arcos, Cecilia Ramirez-Santillán, Concepción Matesanz, María Teresa Portoles, and María Vallet-Regí. 2011. "Covalently Bonded Dendrimer-Maghemite Nanosystems: Nonviral Vectors for In Vitro Gene Magnetofection." *Journal of Materials Chemistry* 21 (12):4598–4604.

Gottesman, Michael M., Tito Fojo, and Susan E. Bates. 2002. "Multidrug Resistance in Cancer: Role of ATP-Dependent Transporters." *Nature Reviews Cancer* 2 (1):48–58.

Gratton, Stephanie E. A., Patricia A. Ropp, Patrick D. Pohlhaus, J. Christopher Luft, Victoria J. Madden, Mary E. Napier, and Joseph M. DeSimone. 2008. "The Effect of Particle Design on Cellular Internalization Pathways." *Proceedings of the National Academy of Sciences* 105 (33):11613–18.

Grün, Michael, Iris Lauer, and Klaus K. Unger. 1997. "The Synthesis of Micrometer- and Submicrometer-Size Spheres of Ordered Mesoporous Oxide MCM-41." *Advanced Materials* 9 (3):254–57.

Gu, Jinlou, Wei Fan, Atsushi Shimojima, and Tatsuya Okubo. 2007. "Organic-Inorganic Mesoporous Nanocarriers Integrated with Biogenic Ligands." *Small* 3 (10):1740–44.

Haag, Rainer, and Felix Kratz. 2006. "Polymer Therapeutics: Concepts and Applications." *Angewandte Chemie International Edition* 45 (8):1198–1215.

Han, Xiao, Yihua Zhu, Xiaoling Yang, Jianmei Zhang, and Chunzhong Li. 2011. "Dendrimer-Encapsulated Pt Nanoparticles on Mesoporous Silica for Glucose Detection." *Journal of Solid State Electrochemistry* 15 (3):511–17.

Hannon, Gregory J., and John J. Rossi. 2004. "Unlocking the Potential of the Human Genome with RNA Interference." *Nature* 431 (7006):371–78.

Harper, G. R., M. C. Davies, S. S. Davis, T. F. Tadros, D. C. Taylor, M. P. Irving, and J. A Waters. 1991. "Steric Stabilization of Microspheres with Grafted Polyethylene Oxide Reduces Phagocytosis by Rat Kupffer Cells In Vitro." *Biomaterials* 12 (7):695–700.

Hauser, H., D. Spitzer, E. Verhoeyen, J. Unsinger, and D. Wirth. 2000. "New Approaches towards Ex Vivo and In Vivo Gene Therapy." *Cells Tissues Organs* 167 (2–3):75–80.

He, Qianjun, and Jianlin Shi. 2011. "Mesoporous Silica Nanoparticle Based Nano Drug Delivery Systems: Synthesis, Controlled Drug Release and Delivery, Pharmacokinetics and Biocompatibility." *Journal of Materials Chemistry* 21 (16):5845–55.

He, Qianjun, Jiamin Zhang, Jianlin Shi, Ziyan Zhu, Linxia Zhang, Wenbo Bu, Limin Guo, and Yu Chen. 2010. "The Effect of PEGylation of Mesoporous Silica Nanoparticles on Nonspecific Binding of Serum Proteins and Cellular Responses." *Biomaterials* 31 (6):1085–92.

He, Qianjun, Zhiwen Zhang, Fang Gao, Yaping Li, and Jianlin Shi. 2011. "In Vivo Biodistribution and Urinary Excretion of Mesoporous Silica Nanoparticles: Effects of Particle Size and PEGylation." *Small* 7 (2):271–80.

Hocine, Ouahiba, Magali Gary-Bobo, David Brevet, Marie Maynadier, Simon Fontanel, Laurence Raehm, Sébastien Richeter, Bernard Loock, Pierre Couleaud, Céline Frochot, Clarence Charnay, Gaëlle Derrien, Monique Smaïhi, Amar Sahmoune, Alain Morère, Philippe Maillard, Marcel Garcia, and Jean-Olivier Durand. 2010. "Silicalites and Mesoporous Silica Nanoparticles for Photodynamic Therapy." *International Journal of Pharmaceutics* 402 (1–2):221–30.

Hoffmann, Frank, Maximilian Cornelius, Jürgen Morell, and Michael Fröba. 2006. "Silica-Based Mesoporous Organic-Inorganic Hybrid Materials. *Angewandte Chemie International Edition* 45 (20):3216–51.

Hoffmann, Frank, and Michael Froba. 2011. "Vitalising Porous Inorganic Silica Networks with Organic Functions-PMOs and Related Hybrid Materials." *Chemical Society Reviews* 40 (2):608–20.

Hom, Christopher, Jie Lu, Monty Liong, Hanzhi Luo, Zongxi Li, Jeffrey I. Zink, and Fuyuhiko Tamanoi. 2010. "Mesoporous Silica Nanoparticles Facilitate Delivery of siRNA to Shutdown Signaling Pathways In Mammalian Cells." *Small* 6 (11):1185–90.

Hsiao, Jong-Kai, Chih-Pin Tsai, Tsai-Hua Chung, Yann Hung, Ming Yao, Hon-Man Liu, Chung-Yuan Mou, Chung-Shi Yang, Yao-Chang Chen, and Dong-Ming Huang. 2008. "Mesoporous Silica Nanoparticles as a Delivery System of Gadolinium for Effective Human Stem Cell Tracking." *Small* 4 (9):1445–52.

Huang, I-Ping, Shu-Pin Sun, Shih-Hsun Cheng, Chia-Hung Lee, Chia-Yan Wu, Chung-Shi Yang, Leu-Wei Lo, and Yiu-Kay Lai. 2011. "Enhanced Chemotherapy of Cancer Using pH-Sensitive Mesoporous Silica Nanoparticles to Antagonize P-glycoprotein-Mediated Drug Resistance." *Molecular Cancer Therapeutics* 10 (5):761–69.

Huang, Xinglu, Linlin Li, Tianlong Liu, Nanjing Hao, Huiyu Liu, Dong Chen, and Fangqiong Tang. 2011. "The Shape Effect of Mesoporous Silica Nanoparticles on Biodistribution, Clearance, and Biocompatibility In Vivo." *ACS Nano* 5 (7):5390–99.

Hudson, Sarah P., Robert F. Padera, Robert Langer, and Daniel S. Kohane. 2008. "The Biocompatibility of Mesoporous Silicates." *Biomaterials* 29 (30):4045–55.

Huh, Seong, Jerzy W. Wiench, Ji-Chul Yoo, Marek Pruski, and Victor S. Y. Lin. 2003. "Organic Functionalization and Morphology Control of Mesoporous Silicas via a Co-condensation Synthesis Method." *Chemistry of Materials* 15 (22):4247–56.

Jain, Rakesh K. 1998. "Delivery of Molecular and Cellular Medicine to Solid Tumors." *Journal of Controlled Release* 53 (1–3):49–67.

Jianquan, Fan, Fang Gang, Wang Xiaodan, Zeng Fang, Xiang Yufei, and Wu Shuizhu. 2011. "Targeted Anticancer Prodrug with Mesoporous Silica Nanoparticles as Vehicles." *Nanotechnology* 22 (45):455102. doi:10.1088/0957-4484/22/45/455102

Juarranz, Ángeles, Pedro Jaén, Francisco Sanz-Rodríguez, Jesús Cuevas, and Salvador González. 2008. "Photodynamic Therapy of Cancer: Basic Principles and Applications." *Clinical and Translational Oncology* 10 (3):148–54.

Kamphuis, Marloes M. J., Angus P. R. Johnston, Georgina K. Such, Henk H. Dam, Richard A. Evans, Andrew M. Scott, Edouard C. Nice, Joan K. Heath, and Frank Caruso. 2010. "Targeting of Cancer Cells Using Click-Functionalized Khandare, Jayant, Marcelo Calderon, Nilesh M. Dagia, and Rainer Haag. 2012. "Multifunctional Dendritic Polymers in Nanomedicine: Opportunities and Challenges." *Chemical Society Reviews* 41 (7):2824–48.

Khemtong, Chalermchai, Chase W. Kessinger, and Jinming Gao. 2009. "Polymeric Nanomedicine for Cancer MR Imaging and Drug Delivery." *Chemical Communications* (24):3497–3510.

Kim, Jaeyun, Hoe Suk Kim, Nohyun Lee, Taeho Kim, Hyoungsu Kim, Taekyung Yu, In Chan Song, Woo Kyung Moon, and Taeghwan Hyeon. 2008. "Multifunctional Uniform Nanoparticles Composed of a Magnetite Nanocrystal Core and a Mesoporous Silica Shell for Magnetic Resonance and Fluorescence Imaging and for Drug Delivery." *Angewandte Chemie International Edition* 47 (44):8438–41.

Kim, Moon Il, Jongmin Shim, Taihua Li, Jinwoo Lee, and Hyun Gyu Park. 2011. "Fabrication of Nanoporous Nanocomposites Entrapping Fe3O4 Magnetic Nanoparticles and Oxidases for Colorimetric Biosensing." *Chemistry—A European Journal* 17 (38):10700–707.

Klink, Daniel, Dirk Schindelhauer, Andreas Laner, Torry Tucker, Zsuzsanna Bebok, Erik M. Schwiebert, A. Christopher Boyd, and Bob J. Scholte. 2004. "Gene Delivery Systems—Gene Therapy Vectors for Cystic Fibrosis." *Journal of Cystic Fibrosis* 3 (Suppl 2):203–12.

Kresge, C. T., M. E. Leonowicz, W. J. Roth, J. C. Vartuli, and J. S. Beck. 1992. "Ordered Mesoporous Molecular Sieves Synthesized by a Liquid-Crystal Template Mechanism." *Nature* 359 (6397):710–12.

Labhasetwar, Vinod. 2005. "Nanotechnology for Drug and Gene Therapy: The Importance of Understanding Molecular Mechanisms of Delivery." *Current Opinion in Biotechnology* 16 (6):674–80.

Lammers, Twan, Silvio Aime, Wim E. Hennink, Gert Storm, and Fabian Kiessling. 2011. "Theranostic Nanomedicine." *Accounts of Chemical Research* 44 (10):1029–38.

Lang, Natacha, and Alain Tuel. 2004. "A Fast and Efficient Ion-Exchange Procedure to Remove Surfactant Molecules from MCM-41 Materials." *Chemistry of Materials* 16 (10):1961–66.

Langer, R. 1998. "Drug Delivery and Targeting." *Nature* 392 (6679 Suppl):5–10.

Lebret, Valérie, Laurence Raehm, Jean-Olivier Durand, Monique Smaïhi, Martinus Werts, Mireille Blanchard-Desce, Delphine Méthy-Gonnod, and Catherine Dubernet. 2008. "Surface Functionalization of Two-Photon Dye-Doped Mesoporous Silica Nanoparticles with Folic Acid: Cytotoxicity Studies with HeLa and MCF-7 Cancer Cells." *Journal of Sol-Gel Science and Technology* 48 (1):32–39.

Lechardeur, Delphine, A. S. Verkman, and Gergely L. Lukacs. 2005. "Intracellular Routing of Plasmid DNA during Non-viral Gene Transfer." *Advanced Drug Delivery Reviews* 57 (5):755–67.

Lee, Chia-Hung, Shih-Hsun Cheng, I. Ping Huang, Jeffrey S. Souris, Chung-Shi Yang, Chung-Yuan Mou, and Leu-Wei Lo. 2010. "Intracellular pH-Responsive Mesoporous Silica Nanoparticles for the Controlled Release of Anticancer Chemotherapeutics." *Angewandte Chemie International Edition* 49 (44):8214–19.

Lee, Chia-Hung, Shih-Hsun Cheng, Yu-Jing Wang, Yu-Ching Chen, Nai-Tzu Chen, Jeffrey Souris, Chin-Tu Chen, Chung-Yuan Mou, Chung-Shi Yang, and Leu-Wei Lo. 2009. "Near-Infrared Mesoporous Silica Nanoparticles for Optical Imaging: Characterization and In Vivo Biodistribution." *Advanced Functional Materials* 19 (2):215–22.

Lee, Ji Eun, Dong Jun Lee, Nohyun Lee, Byung Hyo Kim, Seung Hong Choi, and Taeghwan Hyeon. 2011. "Multifunctional Mesoporous Silica Nanocomposite Nanoparticles for pH Controlled Drug Release and Dual Modal Imaging." *Journal of Materials Chemistry* 21 (42):16869–72.

Lee, Ji Eun, Nohyun Lee, Hyoungsu Kim, Jaeyun Kim, Seung Hong Choi, Jeong Hyun Kim, Taeho Kim, In Chan Song, Seung Pyo Park, Woo Kyung Moon, and Taeghwan Hyeon. 2009. "Uniform Mesoporous Dye-Doped Silica Nanoparticles Decorated with Multiple Magnetite Nanocrystals for Simultaneous Enhanced Magnetic Resonance Imaging, Fluorescence Imaging, and Drug Delivery." *Journal of the American Chemical Society* 132 (2):552–57.

Lee, Ji Eun, Nohyun Lee, Taeho Kim, Jaeyun Kim, and Taeghwan Hyeon. 2011. "Multifunctional Mesoporous Silica Nanocomposite Nanoparticles for Theranostic Applications." *Accounts of Chemical Research* 44 (10):893–902.

Li, He, Jing He, Yanfang Zhao, Dan Wu, Yanyan Cai, Qin Wei, and Minghui Yang. 2011. "Immobilization of Glucose Oxidase and Platinum on Mesoporous Silica Nanoparticles for the Fabrication of Glucose Biosensor." *Electrochimica Acta* 56 (7):2960–65.

Li, S. A., H. A. Liu, L. Li, N. Q. Luo, R. H. Cao, D. H. Chen, and Y. Z. Shao. 2011. "Mesoporous Silica Nanoparticles Encapsulating Gd_2O_3 as a Highly Efficient Magnetic Resonance Imaging Contrast Agent." *Applied Physics Letters* 98 (9). 093704 (3 pages) doi:10.1063/1.3560451

Li, Zongxi, Jonathan C. Barnes, Aleksandr Bosoy, J. Fraser Stoddart, and Jeffrey I. Zink. 2012. "Mesoporous Silica Nanoparticles in Biomedical Applications." *Chemical Society Reviews* 41 (7):2590–2605.

Lim, Myong H., and Andreas Stein. 1999. "Comparative Studies of Grafting and Direct Syntheses of Inorganic-Organic Hybrid Mesoporous Materials." *Chemistry of Materials* 11 (11):3285–95.

Lima, Raquel T., Luis M. Martins, Jose E. Guimaraes, Clara Sambade, and M. Helena Vasconcelos. 2004. "Specific Downregulation of bcl-2 and xIAP by RNAi Enhances the Effects of Chemotherapeutic Agents in MCF-7 Human Breast Cancer Cells." *Cancer Gene Therapy* 11 (5):309–16.

Lin, Hong-Ping, Lu-Yi Yang, Chung-Yuan Mou, Shang-Bin Liu, and Huang-Kuei Lee. 2000. "A Direct Surface Silyl Modification of Acid-Synthesized Mesoporous Silica." *New Journal of Chemistry* 24 (5):253–55.

Lin, Yu-Shen, and Christy L. Haynes. 2010. "Impacts of Mesoporous Silica Nanoparticle Size, pore Ordering, and Pore Integrity on Hemolytic Activity." *Journal of the American Chemical Society* 132 (13):4834–42.

Lin, Yu-Shen, Yann Hung, Jen-Kuan Su, Rain Lee, Chen Chang, Meng-Liang Lin, and Chung-Yuan Mou. 2004. "Gadolinium(III)-Incorporated Nanosized Mesoporous Silica as Potential Magnetic Resonance Imaging Contrast Agents." *The Journal of Physical Chemistry B* 108 (40):15608–11.

Lin, Yu-Shen, Katie R. Hurley, and Christy L. Haynes. 2012. "Critical Considerations in the Biomedical Use of Mesoporous Silica Nanoparticles." *The Journal of Physical Chemistry Letters* 3 (3):364–74.

Lin, Yu-Shen, Chih-Pin Tsai, Hsing-Yi Huang, Chieh-Ti Kuo, Yann Hung, Dong-Ming Huang, Yao-Chang Chen, and Chung-Yuan Mou. 2005. "Well-Ordered Mesoporous Silica Nanoparticles as Cell Markers." *Chemistry of Materials* 17 (18):4570–73.

Lin, Yu-Shen, Si-Han Wu, Yann Hung, Yi-Hsin Chou, Chen Chang, Meng-Liang Lin, Chih-Pin Tsai, and Chung-Yuan Mou. 2006. "Multifunctional Composite Nanoparticles: Magnetic, Luminescent, and Mesoporous." *Chemistry of Materials* 18 (22):5170–72.

Liong, Monty, Sarah Angelos, Eunshil Choi, Kaushik Patel, J. Fraser Stoddart, and Jeffrey I. Zink. 2009. "Mesostructured Multifunctional Nanoparticles for Imaging and Drug Delivery." *Journal of Materials Chemistry* 19 (35):6251–57.

Liong, Monty, Jie Lu, Michael Kovochich, Tian Xia, Stefan G. Ruehm, Andre E. Nel, Fuyuhiko Tamanoi, and Jeffrey I. Zink. 2008. "Multifunctional Inorganic Nanoparticles for Imaging, Targeting, and Drug Delivery." *ACS Nano* 2 (5):889–96.

Liu, Hon-Man, Si-Han Wu, Chen-Wen Lu, Ming Yao, Jong-Kai Hsiao, Yann Hung, Yu-Shen Lin, Chung-Yuan Mou, Chung-Shi Yang, Dong-Ming Huang, and Yao-Chang Chen. 2008. "Mesoporous Silica Nanoparticles Improve Magnetic Labeling Efficiency in Human Stem Cells." *Small* 4 (5):619–26.

Liu, Juewen, Xingmao Jiang, Carlee Ashley, and C. Jeffrey Brinker. 2009. "Electrostatically Mediated Liposome Fusion and Lipid Exchange with a Nanoparticle-Supported Bilayer for Control of Surface Charge, Drug Containment, and Delivery." *Journal of the American Chemical Society* 131 (22):7567–69.

Liu, Yi-Hsin, Hong-Ping Lin, and Chung-Yuan Mou. 2004. "Direct Method for Surface Silyl Functionalization of Mesoporous Silica." *Langmuir* 20 (8):3231–39.

Low, Philip S., Walter A. Henne, and Derek D. Doorneweerd. 2007. "Discovery and Development of Folic-Acid-Based Receptor Targeting for Imaging and Therapy of Cancer and Inflammatory Diseases." *Accounts of Chemical Research* 41 (1):120–29.

Lu, Jie, Eunshil Choi, Fuyuhiko Tamanoi, and Jeffrey I. Zink. 2008. "Light-Activated Nanoimpeller-Controlled Drug Release in Cancer Cells." *Small* 4 (4):421–26.

Lu, Jie, Zongxi Li, Jeffrey I. Zink, and Fuyuhiko Tamanoi. 2012. "In Vivo Tumor Suppression Efficacy of Mesoporous Silica Nanoparticles-Based Drug-Delivery System: Enhanced Efficacy by Folate Modification." *Nanomedicine: Nanotechnology, Biology and Medicine* 8 (2):212–20.

Lu, Jie, Monty Liong, Zongxi Li, Jeffrey I. Zink, and Fuyuhiko Tamanoi. 2010. "Biocompatibility, Biodistribution, and Drug-Delivery Efficiency of Mesoporous Silica Nanoparticles for Cancer Therapy in Animals." *Small* 6 (16):1794–1805.

Lu, Jie, Monty Liong, Sean Sherman, Tian Xia, Michael Kovochich, Andre Nel, Jeffrey Zink, and Fuyuhiko Tamanoi. 2007. "Mesoporous Silica Nanoparticles for Cancer Therapy: Energy-Dependent Cellular Uptake and Delivery of Paclitaxel to Cancer Cells." *NanoBioTechnology* 3 (2):89–95.

Lu, Jie, Monty Liong, Jeffrey I. Zink, and Fuyuhiko Tamanoi. 2007. "Mesoporous Silica Nanoparticles as a Delivery System for Hydrophobic Anticancer Drugs." *Small* 3 (8):1341–46.

Lu, Yunfeng, Hongyou Fan, Aaron Stump, Timothy L. Ward, Thomas Rieker, and C. Jeffrey Brinker. 1999. "Aerosol-Assisted Self-Assembly of Mesostructured Spherical Nanoparticles." *Nature* 398 (6724):223–26.

Luo, Zhong, Kaiyong Cai, Yan Hu, Li Zhao, Peng Liu, Lin Duan, and Weihu Yang. 2011. "Mesoporous Silica Nanoparticles End-Capped with Collagen: Redox-Responsive Nanoreservoirs for Targeted Drug Delivery." *Angewandte Chemie International Edition* 50 (3):640–43.

Ma, Yujie, Roeland J. M. Nolte, and Jeroen J. L. M. Cornelissen. 2012. "Virus-Based Nanocarriers for Drug Delivery." *Advanced Drug Delivery Reviews* 64 (9):811–25.

Maeda, H., G. Y. Bharate, and J. Daruwalla. 2009. "Polymeric Drugs for Efficient Tumor-Targeted Drug Delivery Based on EPR-Effect." *European Journal of Pharmaceutics and Biopharmaceutics* 71 (3):409–19.

Mancuso, Katherine, William W. Hauswirth, Qiuhong Li, Thomas B. Connor, James A. Kuchenbecker, Matthew C. Mauck, Jay Neitz, and Maureen Neitz. 2009. "Gene Therapy for Red-Green Colour Blindness in Adult Primates." *Nature* 461 (7265):784–87.

Martín-Saavedra, F. M., E. Ruíz-Hernández, A. Boré, D. Arcos, M. Vallet-Regí, and N. Vilaboa. 2010. "Magnetic Mesoporous Silica Spheres for Hyperthermia Therapy." *Acta Biomaterialia* 6 (12):4522–31.

Meng, Huan, Monty Liong, Tian Xia, Zongxi Li, Zhaoxia Ji, Jeffrey I. Zink, and Andre E. Nel. 2010. "Engineered Design of Mesoporous Silica Nanoparticles to Deliver Doxorubicin and P-glycoprotein siRNA to Overcome Drug Resistance in a Cancer Cell Line." *ACS Nano* 4 (8):4539–50.

Meng, Huan, Min Xue, Tian Xia, Zhaoxia Ji, Derrick Y. Tarn, Jeffrey I. Zink, and Andre E. Nel. 2011. "Use of Size and a Copolymer Design Feature to Improve the Biodistribution and the Enhanced Permeability and Retention Effect of Doxorubicin-Loaded Mesoporous Silica Nanoparticles in a Murine Xenograft Tumor Model." *ACS Nano* 5 (5):4131–44.

Mintzer, Meredith A., and Eric E. Simanek. 2008. "Nonviral Vectors for Gene Delivery." *Chemical Reviews* 109 (2):259–302.

Moazed, Danesh. 2009. "Small RNAs in Transcriptional Gene Silencing and Genome Defence." *Nature* 457 (7228):413–20.

Morille, Marie, Catherine Passirani, Sandrine Dufort, Guillaume Bastiat, Bruno Pitard, Jean-Luc Coll, and Jean-Pierre Benoit. 2011. "Tumor Transfection after Systemic Injection of DNA Lipid Nanocapsules." *Biomaterials* 32 (9):2327–33.

Mortera, Renato, Juan Vivero-Escoto, Igor I. Slowing, Edoardo Garrone, Barbara Onida, and Victor S. Y. Lin. 2009. "Cell-Induced Intracellular Controlled Release of Membrane Impermeable Cysteine from a Mesoporous Silica Nanoparticle-Based Drug Delivery System." *Chemical Communications* (22):3219–21.

Muggia, Franco M. 1999. "Doxorubicin-Polymer Conjugates: Further Demonstration of the Concept of Enhanced Permeability and Retention." *Clinical Cancer Research* 5 (1):7–8.

Mulder, Willem J. M., Gustav J. Strijkers, Geralda A. F. van Tilborg, David P. Cormode, Zahi A. Fayad, and Klaas Nicolay. 2009. "ChemInform Abstract: Nanoparticulate Assemblies of Amphiphiles and Diagnostically Active Materials for Multimodality Imaging." *ChemInform* 40 (47). doi:10.1002/chun.200947269

Na, Hyon Bin, In Chan Song, and Taeghwan Hyeon. 2009. "Inorganic Nanoparticles for MRI Contrast Agents." *Advanced Materials* 21 (21):2133–48.

Neu, Michael, Dagmar Fischer, and Thomas Kissel. 2005. "Recent Advances in Rational Gene Transfer Vector Design Based on Poly(ethylene imine) and Its Derivatives." *The Journal of Gene Medicine* 7 (8):992–1009.

Oh, Kyung T., Haiqing Yin, Eun Seong Lee, and You Han Bae. 2007. "Polymeric Nanovehicles for Anticancer Drugs with Triggering Release Mechanisms." *Journal of Materials Chemistry* 17 (38):3987–4001.

Orkin, Stuart H. 1986. "Molecular Genetics and Potential Gene Therapy." *Clinical Immunology and Immunopathology* 40 (1):151–56.

Ortel, B., C. R. Shea, and P. Calzavara-Pinton. 2009. "Molecular Mechanisms of Photodynamic Therapy." *Frontiers in Bioscience* 14:4157–72.

Owens Iii, Donald E., and Nicholas A. Peppas. 2006. "Opsonization, Biodistribution, and Pharmacokinetics of Polymeric Nanoparticles." *International Journal of Pharmaceutics* 307 (1):93–102.

Özalp, Veli Cengiz; Eyidogan, Fusun; Oktem, Huseyin Avni. 2011. "Aptamer-Gated Nanoparticles for Smart Drug Delivery." *Pharmaceuticals* 4 (8):1137–57.

Paillard, Archibald, Catherine Passirani, Patrick Saulnier, Maya Kroubi, Emmanuel Garcion, Jean-Pierre Benoît, and Didier Betbeder. 2010. "Positively-Charged, Porous, Polysaccharide Nanoparticles Loaded with Anionic Molecules Behave as 'Stealth' Cationic Nanocarriers." *Pharmaceutical Research* 27 (1):126–33.

Pakunlu, Refika I., Thomas J. Cook, and Tamara Minko. 2003. "Simultaneous Modulation of Multidrug Resistance and Antiapoptotic Cellular Defense by MDR1 and BCL-2 Targeted Antisense Oligonucleotides Enhances the Anticancer Efficacy of Doxorubicin." *Pharmaceutical Research* 20 (3):351–59.

Pakunlu, Refika I., Yang Wang, William Tsao, Vitaly Pozharov, Thomas J. Cook, and Tamara Minko. 2004. "Enhancement of the Efficacy of Chemotherapy for Lung Cancer by Simultaneous Suppression of Multidrug Resistance and Antiapoptotic Cellular Defense." *Cancer Research* 64 (17):6214–24.

Park, In Young, In Yong Kim, Mi Kyong Yoo, Yun Jae Choi, Myung-Haing Cho, and Chong Su Cho. 2008. "Mannosylated Polyethylenimine Coupled Mesoporous Silica Nanoparticles for Receptor-Mediated Gene Delivery." *International Journal of Pharmaceutics* 359 (1–2):280–87.

Peer, Dan, Jeffrey M. Karp, Seungpyo Hong, Omid C. Farokhzad, Rimona Margalit, and Robert Langer. 2007. "Nanocarriers as an Emerging Platform for Cancer Therapy." *Nature Nanotechnology* 2 (12):751–60.

Piao, Yuanzhe, Andrew Burns, Jaeyun Kim, Ulrich Wiesner, and Taeghwan Hyeon. 2008. "Designed Fabrication of Silica-Based Nanostructured Particle Systems for Nanomedicine Applications." *Advanced Functional Materials* 18 (23):3745–58.

Pinholt, Charlotte, Jens Thostrup Bukrinsky, Susanne Hostrup, Sven Frokjaer, Willem Norde, and Lene Jorgensen. 2011. "Influence of PEGylation with Linear and Branched PEG Chains on the Adsorption of Glucagon to Hydrophobic Surfaces." *European Journal of Pharmaceutics and Biopharmaceutics* 77 (1):139–47.

Pirollo, Kathleen F., and Esther H. Chang. 2008. "Does a Targeting Ligand Influence Nanoparticle Tumor Localization or Uptake?" *Trends in Biotechnology* 26 (10):552–58.

Radu, Daniela R., Cheng-Yu Lai, Ksenija Jeftinija, Eric W. Rowe, Srdija Jeftinija, and Victor S. Y. Lin. 2004. "A Polyamidoamine Dendrimer-Capped Mesoporous Silica Nanosphere-Based Gene Transfection Reagent." *Journal of the American Chemical Society* 126 (41):13216–17.

Radu, Daniela R., Cheng-Yu Lai, Jerzy W. Wiench, Marek Pruski, and Victor S. Y. Lin. 2004. "Gatekeeping Layer Effect: A Poly(lactic acid)-Coated Mesoporous Silica Nanosphere-Based Fluorescence Probe for Detection of Amino-Containing Neurotransmitters." *Journal of the American Chemical Society* 126 (6):1640–41.

Ritter, Hanna, and Dominik Brühwiler. 2009. "Accessibility of Amino Groups in Postsynthetically Modified Mesoporous Silica." *The Journal of Physical Chemistry C* 113 (24):10667–74.

Robertson, C. A., D. Hawkins Evans, and H. Abrahamse. 2009. "Photodynamic Therapy (PDT): A Short Review on Cellular Mechanisms and Cancer Research Applications for PDT." *Journal of Photochemistry and Photobiology B: Biology* 96 (1):1–8.

Rosenholm, Jessica M., and Mika Lindén. 2008. "Towards Establishing Structure-Activity Relationships for Mesoporous Silica in Drug Delivery Applications." *Journal of Controlled Release* 128 (2):157–64.

Rosenholm, Jessica M., Annika Meinander, Emilia Peuhu, Rasmus Niemi, John E. Eriksson, Cecilia Sahlgren, and Mika Lindén. 2008. "Targeting of Porous Hybrid Silica Nanoparticles to Cancer Cells." *ACS Nano* 3 (1):197–206.

Rosenholm, Jessica M., Emilia Peuhu, Laurel Tabe Bate-Eya, John E. Eriksson, Cecilia Sahlgren, and Mika Lindén. 2010. "Cancer-Cell-Specific Induction of Apoptosis Using Mesoporous Silica Nanoparticles as Drug-Delivery Vectors." *Small* 6 (11):1234–41.

Rosenholm, Jessica M., Emilia Peuhu, John E. Eriksson, Cecilia Sahlgren, and Mika Lindén. 2009. "Targeted Intracellular Delivery of Hydrophobic Agents Using Mesoporous Hybrid Silica Nanoparticles as Carrier Systems." *Nano Letters* 9 (9):3308–11.

Rosenholm, Jessica M., Cecilia Sahlgren, and Mika Lindén. 2010. "Towards Multifunctional, Targeted Drug Delivery Systems Using Mesoporous Silica Nanoparticles—Opportunities & Challenges." *Nanoscale* 2 (10):1870–83.

Rosenholm, Jessica M., Cecilia Sahlgren, and Mika Lindén. 2011. "Multifunctional Mesoporous Silica Nanoparticles for Combined Therapeutic, Diagnostic and Targeted Action in Cancer Treatment." *Current Drug Targets* 12 (8):1166–86.

Rosi, Nathaniel L., David A. Giljohann, C. Shad Thaxton, Abigail K. R. Lytton-Jean, Min Su Han, and Chad A. Mirkin. 2006. "Oligonucleotide-Modified Gold Nanoparticles for Intracellular Gene Regulation." *Science* 312 (5776):1027–30.

Rubanyi, Gabor M. 2001. "The Future of Human Gene Therapy." *Molecular Aspects of Medicine* 22 (3):113–42.

Ruiz-Hernández, E., A. López-Noriega, D. Arcos, I. Izquierdo-Barba, O. Terasaki, and M. Vallet-Regí. 2007. "Aerosol-Assisted Synthesis of Magnetic Mesoporous Silica Spheres for Drug Targeting." *Chemistry of Materials* 19 (14):3455–63.

Ruiz-Hernández, E., A. López-Noriega, D. Arcos, and M. Vallet-Regí. 2008. "Mesoporous Magnetic Microspheres for Drug Targeting." *Solid State Sciences* 10 (4):421–26.

Saad, Maha, Olga B. Garbuzenko, and Tamara Minko. 2008. "Co-delivery of siRNA and an Anticancer Drug for Treatment of Multidrug-Resistant Cancer." *Nanomedicine* 3 (6):761–76.

Sadzuka, Yasuyuki, Ikumi Sugiyama, Tomoko Tsuruda, and Takashi Sonobe. 2006. "Characterization and Cytotoxicity of Mixed Polyethyleneglycol Modified Liposomes Containing Doxorubicin." *International Journal of Pharmaceutics* 312 (1–2):83–89.

Saha, S., K. C. F. Leung, T. D. Nguyen, J. F. Stoddart, and J. I. Zink. 2007. "Nanovalves." *Advanced Functional Materials* 17 (5):685–93.

Shi, Jinjun, Alexander R. Votruba, Omid C. Farokhzad, and Robert Langer. 2010. "Nanotechnology in Drug Delivery and Tissue Engineering: From Discovery to Applications." *Nano Letters* 10 (9):3223–30.

Simmchen, Juliane, Alejandro Baeza, Daniel Ruiz, Maria José Esplandiu, and María Vallet-Regí. 2012. "Asymmetric Hybrid Silica Nanomotors for Capture and Cargo Transport: Towards a Novel Motion-Based DNA Sensor." *Small* 8 (13):2053–59. doi:10.1002/smll.201101593

Slowing, I. I., B. G. Trewyn, S. Giri, and V. S. Y. Lin. 2007. "Mesoporous Silica Nanoparticles for Drug Delivery and Biosensing Applications." *Advanced Functional Materials* 17 (8):1225–36.

Slowing, Igor, Brian G. Trewyn, and Victor S. Y. Lin. 2006. "Effect of Surface Functionalization of MCM-41-Type Mesoporous Silica Nanoparticles on the Endocyto.

Slowing, Igor I., Juan L. Vivero-Escoto, Chia-Wen Wu, and Victor S. Y. Lin. 2008. "Mesoporous Silica Nanoparticles as Controlled Release Drug Delivery and Gene Transfection Carriers." *Advanced Drug Delivery Reviews* 60 (11):1278–88.

Smith, Andrew M., Hongwei Duan, Aaron M. Mohs, and Shuming Nie. 2008. "Bioconjugated Quantum Dots for In Vivo Molecular and Cellular Imaging." *Advanced Drug Delivery Reviews* 60 (11):1226–40.

Sokolova, Viktoriya, and Matthias Epple. 2008. "Inorganic Nanoparticles as Carriers of Nucleic Acids into Cells." *Angewandte Chemie International Edition* 47 (8):1382–95.

Soler-Illia, Galo J. A. A., and Omar Azzaroni. 2011. "Multifunctional Hybrids by Combining Ordered Mesoporous Materials and Macromolecular Building Blocks." *Chemical Society Reviews* 40 (2):1107–50.

Souris, Jeffrey S., Chia-Hung Lee, Shih-Hsun Cheng, Chin-Tu Chen, Chung-Shi Yang, Ja-an A. Ho, Chung-Yuan Mou, and Leu-Wei Lo. 2010. "Surface Charge-Mediated Rapid Hepatobiliary Excretion of Mesoporous Silica Nanoparticles." *Biomaterials* 31 (21):5564–74.

Sudimack, Jennifer, and Robert J. Lee. 2000. "Targeted Drug Delivery via the Folate Receptor." *Advanced Drug Delivery Reviews* 41 (2):147–62.

Suh, Won Hyuk, Yoo-Hun Suh, and Galen D. Stucky. 2009. "Multifunctional Nanosystems at the Interface of Physical and Life Sciences." *Nano Today* 4 (1):27–36.

Sun, Conroy, Jerry S. H. Lee, and Miqin Zhang. 2008. "Magnetic Nanoparticles in MR Imaging and Drug Delivery." *Advanced Drug Delivery Reviews* 60 (11):1252–65.

Tan, Weihong, Kemin Wang, Xiaoxiao He, Xiaojun Julia Zhao, Timothy Drake, Lin Wang, and Rahul P. Bagwe. 2004. "Bionanotechnology Based on Silica Nanoparticles." *Medicinal Research Reviews* 24 (5):621–38.

Taylor, Kathryn M. L., Jason S. Kim, William J. Rieter, Hongyu An, Weili Lin, and Wenbin Lin. 2008. "Mesoporous Silica Nanospheres as Highly Efficient MRI Contrast Agents." *Journal of the American Chemical Society* 130 (7):2154–55.

Taylor-Pashow, Kathryn M. L., Joseph Della Rocca, Rachel C. Huxford, and Wenbin Lin. 2010. "Hybrid Nanomaterials for Biomedical Applications." *Chemical Communications* 46 (32):5832–49.

Terreno, Enzo, Daniela Delli Castelli, Alessandra Viale, and Silvio Aime. 2010. "Challenges for Molecular Magnetic Resonance Imaging." *Chemical Reviews* 110 (5):3019–42.

Thomas, Courtney R., Daniel P. Ferris, Jae-Hyun Lee, Eunjoo Choi, Mi Hyeon Cho, Eun Sook Kim, J. Fraser Stoddart, Jeon-Soo Shin, Jinwoo Cheon, and Jeffrey I. Zink. 2010. "Noninvasive Remote-Controlled Release of Drug Molecules In Vitro Using Magnetic Actuation of Mechanized Nanoparticles." *Journal of the American Chemical Society* 132 (31):10623–25.

Torchilin, Vladimir P. 2005. "Recent Advances with Liposomes as Pharmaceutical Carriers." *Nature Reviews Drug Discovery* 4 (2):145–60.

Torchilin, Vladimir. 2011. "Tumor Delivery of Macromolecular Drugs Based on the EPR Effect." *Advanced Drug Delivery Reviews* 63 (3):131–35.

Tourne-Peteilh, Corine, Daniel Brunel, Sylvie Begu, Bich Chiche, Francois Fajula, Dan A. Lerner, and Jean-Marie Devoisselle. 2003. "Synthesis and Characterisation of Ibuprofen-Anchored MCM-41 Silica and Silica Gel." *New Journal of Chemistry* 27 (10):1415–18.

Trewyn, Brian G., Supratim Giri, Igor I. Slowing, and Victor S. Y. Lin. 2007. "Mesoporous Silica Nanoparticle Based Controlled Release, Drug Delivery, and Biosensor Systems." *Chemical Communications* 31:3236–45.

Tsai, Chih-Pin, Chao-Yu Chen, Yann Hung, Fu-Hsiung Chang, and Chung-Yuan Mou. 2009. "Monoclonal Antibody-Functionalized Mesoporous Silica Nanoparticles (MSN) for Selective Targeting Breast Cancer Cells." *Journal of Materials Chemistry* 19 (32):5737–43.

Tseng, Yu-Cheng, Subho Mozumdar, and Leaf Huang. 2009. "Lipid-Based Systemic Delivery of siRNA." *Advanced Drug Delivery Reviews* 61 (9):721–31.

Urban-Klein, B., S. Werth, S. Abuharbeid, F. Czubayko, and A. Aigner. 2004. "RNAi-Mediated Gene-Targeting through Systemic Application of Polyethylenimine (PEI)-Complexed siRNA In Vivo." *Gene Therapy* 12 (5):461–66.

Vakil, Varsha, Joanna J. Sung, Marta Piecychna, Jeffrey R. Crawford, Phillip Kuo, Ali K. Abu-Alfa, Shawn E. Cowper, Richard Bucala, and Richard H. Gomer. 2009. "Gadolinium-Containing Magnetic Resonance Image Contrast Agent Promotes Fibrocyte Differentiation." *Journal of Magnetic Resonance Imaging* 30 (6):1284–88.

Vallet-Regí, M., A. Rámila, R. P. del Real, and J. Pérez-Pariente. 2001. "A New Property of MCM-41: Drug Delivery System." *Chemistry of Materials* 13 (2):308–11.

Vallet-Regí, María, Francisco Balas, and Daniel Arcos. 2007. "Mesoporous Materials for Drug Delivery." *Angewandte Chemie International Edition* 46 (40):7548–58.

Vallet-Regí, María, Montserrat Colilla, and Blanca Gonzalez. 2011. "Medical Applications of Organic-Inorganic Hybrid Materials within the Field of Silica-Based Bioceramics." *Chemical Society Reviews* 40 (2):596–607.

Vallet-Regí, María, and Eduardo Ruiz-Hernández. 2011. "Bioceramics: From Bone Regeneration to Cancer Nanomedicine." *Advanced Materials* 23 (44):5177–5218.

Vallet-Regí, María, Eduardo Ruiz-Hernández, Blanca González, and Alejandro Baeza. 2011. "Design of Smart Nanomaterials for Drug and Gene Delivery." *Journal of Biomaterials and Tissue Engineering* 1 (1):6–29.

van Landeghem, Frank K. H., K. Maier-Hauff, A. Jordan, Karl-Titus Hoffmann, U. Gneveckow, R. Scholz, B. Thiesen, W. Brück, and A. von Deimling. 2009. "Post-Mortem Studies in Glioblastoma Patients Treated with Thermotherapy Using Magnetic Nanoparticles." *Biomaterials* 30 (1):52–57.

van Schooneveld, Matti M., David P. Cormode, Rolf Koole, J. Timon van Wijngaarden, Claudia Calcagno, Torjus Skajaa, Jan Hilhorst, Dannis C. 't Hart, Zahi A. Fayad, Willem J. M. Mulder, and Andries Meijerink. 2010. "A Fluorescent, Paramagnetic and PEGylated Gold/Silica Nanoparticle for MRI, CT and Fluorescence Imaging." *Contrast Media & Molecular Imaging* 5 (4):231–36.

Vauthier, Christine, Nicolas Tsapis, and Patrick Couvreur. 2010. "Nanoparticles: Heating Tumors to Death?" *Nanomedicine* 6 (1):99–109.

Verma, Ayush, Oktay Uzun, Yuhua Hu, Ying Hu, Hee-Sun Han, Nicki Watson, Suelin Chen, Darrell J. Irvine, and Francesco Stellacci. 2008. "Surface-Structure-Regulated Cell-Membrane Penetration by Monolayer-Protected Nanoparticles." *Nature Materials* 7 (7):588–95.

Veronese, Francesco M., and Gianfranco Pasut. 2005. "PEGylation, Successful Approach to Drug Delivery." *Drug Discovery Today* 10 (21):1451–58.

Villaraza, Aaron Joseph L., Ambika Bumb, and Martin W. Brechbiel. 2010. "Macromolecules, Dendrimers, and Nanomaterials in Magnetic Resonance Imaging: The Interplay between Size, Function, and Pharmacokinetics." *Chemical Reviews* 110 (5):2921–59.

Vivero-Escoto, Juan L., Rachel C. Huxford-Phillips, and Wenbin Lin. 2012. "Silica-Based Nanoprobes for Biomedical Imaging and Theranostic Applications." *Chemical Society Reviews* 41 (7):2673–85.

Vivero-Escoto, Juan L., Igor I. Slowing, Brian G. Trewyn, and Victor S. Y. Lin. 2010. "Mesoporous Silica Nanoparticles for Intracellular Controlled Drug Delivery." *Small* 6 (18):1952–67.

Vivero-Escoto, Juan L., Kathryn M. L. Taylor-Pashow, Rachel C. Huxford, Joseph Della Rocca, Christie Okoruwa, Hongyu An, Weili Lin, and Wenbin Lin. 2011. "Multifunctional Mesoporous Silica Nanospheres with Cleavable Gd(III) Chelates as MRI Contrast Agents: Synthesis, Characterization, Target-Specificity, and Renal Clearance." *Small* 7 (24):3519–28.

Vonarbourg, A., C. Passirani, P. Saulnier, P. Simard, J. C. Leroux, and J. P. Benoit. 2006. "Evaluation of Pegylated Lipid Nanocapsules versus Complement System Activation and Macrophage Uptake." *Journal of Biomedical Materials Research Part A* 78A (3):620–28.

Wang, Fang, Xiaolan Chen, Zengxia Zhao, Shaoheng Tang, Xiaoqing Huang, Chenghong Lin, Congbo Cai, and Nanfeng Zheng. 2011. "Synthesis of Magnetic, Fluorescent and Mesoporous Core-Shell-Structured Nanoparticles for Imaging, Targeting and Photodynamic Therapy." *Journal of Materials Chemistry* 21 (30):11244–52.

Wang, Li-Sheng, Li-Chen Wu, Shin-Yi Lu, Li-Ling Chang, I. Ting Teng, Chia-Min Yang, and Ja-an Annie Ho. 2010. "Biofunctionalized Phospholipid-Capped Mesoporous Silica Nanoshuttles for Targeted Drug Delivery: Improved Water Suspensibility and Decreased Nonspecific Protein Binding." *ACS Nano* 4 (8):4371–79.

Wang, Tingting, Lingyu Zhang, Zhongmin Su, Chungang Wang, Yi Liao, and Qin Fu. 2011. "Multifunctional Hollow Mesoporous Silica Nanocages for Cancer Cell Detection and the Combined Chemotherapy and Photodynamic Therapy." *ACS Applied Materials & Interfaces* 3 (7):2479–86.

Wang, Yong, Shujun Gao, Wen-Hui Ye, Ho Sup Yoon, and Yi-Yan Yang. 2006. "Co-delivery of Drugs and DNA from Cationic Core-Shell Nanoparticles Self-Assembled from a Biodegradable Copolymer." *Nature Materials* 5 (10):791–96.

Wei, Yen, Hua Dong, Jigeng Xu, and Qiuwei Feng. 2002. "Simultaneous Immobilization of Horseradish Peroxidase and Glucose Oxidase in Mesoporous Sol-Gel Host Materials." *ChemPhysChem* 3 (9):802–8.

Whitehead, Kathryn A., Robert Langer, and Daniel G. Anderson. 2009. "Knocking Down Barriers: Advances in siRNA Delivery." *Nature Reviews Drug Discovery* 8 (2):129–38.

Woodrow, Kim A., Yen Cu, Carmen J. Booth, Jennifer K. Saucier-Sawyer, Monica J. Wood, and W. Mark Saltzman. 2009. "Intravaginal Gene Silencing Using Biodegradable Polymer Nanoparticles Densely Loaded with Small-Interfering RNA." *Nature Materials* 8 (6):526–33.

Wu, Si-Han, Yann Hung, and Chung-Yuan Mou. 2011. "Mesoporous Silica Nanoparticles as Nanocarriers." *Chemical Communications* 47 (36):9972–85.

Xia, Tian, Michael Kovochich, Jonathan Brant, Matt Hotze, Joan Sempf, Terry Oberley, Constantinos Sioutas, Joanne I. Yeh, Mark R. Wiesner, and Andre E. Nel. 2006. "Comparison of the Abilities of Ambient and Manufactured Nanoparticles to Induce Cellular Toxicity According to an Oxidative Stress Paradigm." *Nano Letters* 6 (8):1794–1807.

Xia, Tian, Michael Kovochich, Monty Liong, Huan Meng, Sanaz Kabehie, Saji George, Jeffrey I. Zink, and Andre E. Nel. 2009. "Polyethyleneimine Coating Enhances the Cellular Uptake of Mesoporous Silica Nanoparticles and Allows Safe Delivery of siRNA and DNA Constructs." *ACS Nano* 3 (10):3273–86.

Xia, Tian, Michael Kovochich, Monty Liong, Jeffrey I. Zink, and Andre E. Nel. 2007. "Cationic Polystyrene Nanosphere Toxicity Depends on Cell-Specific Endocytic and Mitochondrial Injury Pathways." *ACS Nano* 2 (1):85–96.

Yano, Junichi, Kazuko Hirabayashi, Shin-ichiro Nakagawa, Tohru Yamaguchi, Masaki Nogawa, Isao Kashimori, Haruna Naito, Hidetoshi Kitagawa, Kouichi Ishiyama, Tadaaki Ohgi, and Tatsuro Irimura. 2004. "Antitumor Activity of Small Interfering RNA/ Cationic Liposome Complex in Mouse Models of Cancer." *Clinical Cancer Research* 10 (22):7721–26.

Zapilko, Clemens, Markus Widenmeyer, Iris Nagl, Frank Estler, Reiner Anwander, Gabriele Raudaschl-Sieber, Olaf Groeger, and Günter Engelhardt. 2006. "Advanced Surface Functionalization of Periodic Mesoporous Silica: Kinetic Control by Trisilazane Reagents." *Journal of the American Chemical Society* 128 (50):16266–76.

Zeisig, Reiner, Kazuhiko Shimada, Sadao Hirota, and Dieter Arndt. 1996. "Effect of Sterical Stabilization on Macrophage Uptake In Vitro and on Thickness of the Fixed Aqueous Layer of Liposomes Made from Alkylphosphocholines." *Biochimica et Biophysica Acta (BBA): Biomembranes* 1285 (2):237–45.

Zhao, Yannan, Juan L. Vivero-Escoto, Igor I. Slowing, Brian G. Trewyn, and Victor S.-Y. Lin. 2010. "Capped Mesoporous Silica Nanoparticles as Stimuli-Responsive Controlled Release Systems for Intracellular Drug/Gene Delivery." *Expert Opinion on Drug Delivery* 7 (9):1013–29.

Zhu, Chun-Ling, Xue-Yuan Song, Wen-Hui Zhou, Huang-Hao Yang, Yong-Hong Wen, and Xiao-Ru Wang. 2009. "An Efficient Cell-Targeting and Intracellular Controlled-Release Drug Delivery System Based on MSN-PEM-Aptamer Conjugates." *Journal of Materials Chemistry* 19 (41):7765–70.

5 Bone Tissue Regeneration

Trauma, disease, and injury that cause damage and degradation of tissues have powered the development of treatments to facilitate their repair, replacement, or regeneration (O'Brien 2011). Traditionally, the treatment has been based on: (1) transplanting healthy tissue from the same patient to the damaged area (autograft), and (2) transplanting healthy tissue from a donor to the patient (allograft).

Although those approaches have been a great solution within the last few years, there are major drawbacks associated with both techniques. The use of autografts is painful, restricted to anatomical limitations, and, more importantly, related to donor-site morbidity because of possible infection and hematoma. On the other hand, allografts also present the limitation of accessing enough tissue for all the patients, the risk of rejection by the immune system of the patient, and the chance of infection together with the possibility of a disease contagion from the donor to the patient.

The developing field of tissue engineering targets the treatment and repair of damaged tissues with a different approach. Instead of replacing them, tissue engineering aims to regenerate damaged tissues by developing biological substitutes that restore, maintain, or improve tissue function (Langer 2000).

Nowadays, ordered mesoporous ceramics have not yet been applied as pure tissue-engineering constructs. They have been successfully employed as bone tissue regeneration materials, which is the previous stage before using them as tissue-engineering scaffolds. In this way, ordered mesoporous ceramics have been applied, both in vitro and in vivo, as second-generation materials. As a consequence of the positive results obtained, as will be shown in this chapter, the next step in the near future would be employing ordered mesoporous materials to fabricate 3D scaffolds for tissue engineering.

However, although ordered mesoporous ceramics are still under research to be used as tissue-engineering constructs, the authors of this book would like to define some concepts of tissue engineering, so the reader will easily understand the needs in the future research of mesoporous ceramics for tissue engineering.

5.1 TISSUE ENGINEERING

Tissue-engineering technology is based on the combination of cells from the body with scaffolds that act as templates for tissue regeneration to guide and promote the growth of new tissue. The concept of *tissue engineering* was officially defined at a National Science Foundation in 1988 as *the application of principles and methods of engineering and life sciences toward the fundamental understanding of structure-function relationships in normal and pathological mammalian tissues and the development of biological substitutes to restore, maintain, or improve tissue function* (Vallet-Regí and Ruiz-Hernández 2011). The research area of tissue engineering

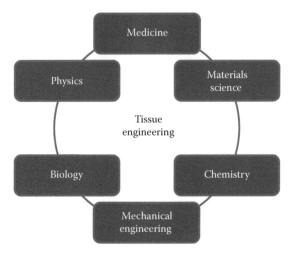

FIGURE 5.1 The knowledge from biology, cellular and molecular biology, and biochemistry is needed for the design of the new tissues. At the same time, chemistry together with materials science and engineering bring the required knowledge for designing and building the scaffolds in which cells should attach and grow. Finally, the knowledge from medicine applies practical issues to real problems and necessities.

is a highly multidisciplinary field, with experts from medicine, materials science, mechanical engineering, biology, chemistry, and physics, as schematically represented in Figure 5.1

In the fabrication of novel tissues, there are three basic elements that need to be carefully designed and/or selected: an appropriate scaffold, enough cells attached to those scaffolds, and specific signals to trigger the formation of the desired tissue, as shown in Figure 5.2.

The appropriate selection of each of those elements will depend on the type of tissue that needs to be repaired. The material to build the scaffold will have to be selected to fulfill the main role of the scaffold: direct the cell growth that will form the new tissue. Thus this material should be an adequate substrate that permits cellular adhesion and proliferation, and allows the cells to perform their functions. The experience in the clinic shows that normally in natural regeneration processes, the presence of biological signals is basic to directing, organizing, and promoting that regeneration. This is the reason that specific signals in the scaffolds need to be implanted, to be then released at the specific site to trigger the regeneration process. Thus the materials selected for the scaffold fabrication should present certain drug delivery capabilities.

In this chapter we will focus on bone tissue engineering because it is where bioceramic materials can realistically be employed, as has been done in the last few years. Thus before entering onto considerations on which materials are more convenient for the scaffold production, a quick overview of bone tissue engineering will be presented.

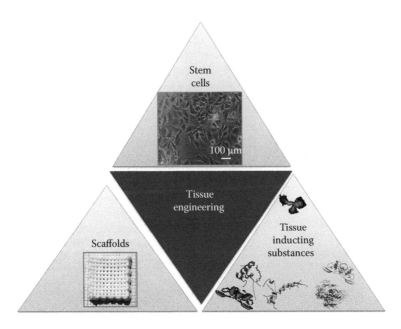

FIGURE 5.2   Three basic elements that must be carefully designed within tissue engineering technologies.

5.2   BONE TISSUE ENGINEERING

Considering the definition of tissue engineering published by Langer and Vacanti (Langer and Vacanti 1993), bone tissue engineering can be defined as *an emerging interdisciplinary field that seeks to address the needs by applying the principles of biology and engineering to the development of viable substitutes that restore and maintain the function of human bone tissues.* Thus the real challenge in bone tissue engineering is to mimic nature's behavior (Place, Evans, and Stevens 2009). It is for this reason that is necessary to understand the hierarchical structure of bone (Figure 5.3) before starting to design any scaffold for bone tissue engineering.

5.2.1   BONE TISSUE

Bone tissue is a natural composite material made of a combination of organic and inorganic components:

- Collagen (organic biopolymer) is a triple helix of protein chains that presents high tensile and flexural strength. The role of this biopolymer is basically to provide a framework for bone tissue.
- Carbonate hydroxyapatite (inorganic ceramic) is a crystalline calcium phosphate that provides the stiffness and high compressive strength of bone (Vallet-Regí and Arcos 2008).

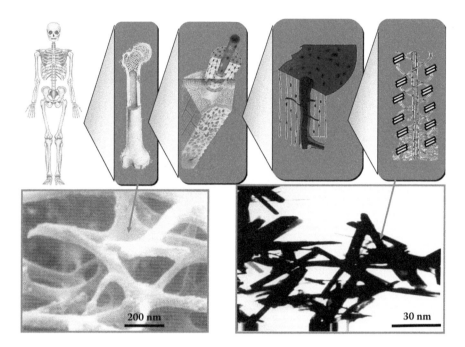

FIGURE 5.3 (**See color insert.**) Representation of the hierarchical structure of bone (top); and micrographs of bone macroporosity (bottom left) and carbonate hydroxyapatite (bottom right).

Additionally, it has been observed that there are two types of bone in vertebrates:

- Cortical bone (also called compact bone) is a dense structure with high mechanical strength.
- Cancellous bone (also called trabecular or spongy bone) is less dense and weaker than cortical bone because of its porous nature. As a consequence of its porous structure, it is highly vascularized and contains red bone marrow to produce blood cells.

Bone plays very important roles in critical functions in human physiology, such as protection, movement, and support of other critical organs, blood production, mineral storage and homeostasis, blood pH regulation, multiple progenitor cell housing, and others (Porter, Ruckh, and Popat 2009).

5.2.2 BONE REMODELING

Bone is dynamic tissue that suffers continuous changes in structure and shape in response to local loading, which gives the skeleton its regeneration capacity and functional adaptation. This dynamic process is called bone remodeling, and it is regulated by genetic, mechanical, vascular, hormonal, and local factors. The living cells responding to those factors and in charge of performing the bone remodeling are osteoblasts and osteoclasts. Osteoblasts produce the synthesis of bone matrix and, subsequently, bone formation. On the other hand, osteoclasts degrade the mineralized matrix and therefore produce bone resorption of the old bone tissue that is not required. In a healthy situation, there is normally a balance between bone gained and bone lost. However, in some metabolic bone diseases such as osteoporosis this equilibrium is disrupted and there is an increased bone remodeling due to a high bone resorption, which might lead to an increased fracture risk.

5.2.3 Bone Defects

In case of minor damage in osseous tissue, bone is capable of fully self-repairing through the reactivation of the processes that take place during embryogenesis. In this way, when there is an osseous injury the osteogenesis mechanisms repair the bone tissue at the site of the injury. Normally, the dynamic remodeling mechanisms are good enough to repair the common defects.

However, in case of massive loss of osseous tissue or when the defect exceeds a critical size, as a result of diseased tissue resection as a consequence of bone cancer, bone tissue alone is not able to self-repair. In those cases, the traditional solution has consisted of employing bone substitutes, such as graft implants and synthetic bone materials, to fill those bone defects. From a more modern tissue engineering perspective, the implanted material not only should act as a bone defect filler but also should regenerate and repair the osseous tissues. The employed materials are normally scaffolds that can guide and stimulate bone growth in the approach of regenerative medicine (Figure 5.4). In fact, the latest trends in bone tissue engineering are focused on tissue regeneration rather than on traditional tissue substitution (Vallet-Regí and Ruiz-Hernández 2011).

FIGURE 5.4 Photograph of a bioceramic scaffold (top) and clinical situation to implant the same scaffold in maxillofacial surgery.

According to this modern vision of tissue engineering, the first step of this approach consists of harvesting osteogenic cells from the patient. Once these cells have been expanded in culture, they are seeded on a scaffold that will act as template and stimulus for tissue growth in three dimensions. It is also possible to add tissue-inducing substances such as certain peptides, hormones, and growth factors. Subsequently, the osteogenic cells will express a collagen-enriched extracellular matrix that will calcify. At this point, the scaffold will be implanted into the patient, and after some time it will be resorbed, allowing the newly formed bone to remodel into mature bone.

A variation to this approach consists of chemically grafting or physically adsorbing the previously mentioned tissue-inducing substances (peptides, hormones, or growth factors) to the scaffold itself, which will be implanted into the patient. In this strategy, the osteogenic factors will induce cells from the patient to regenerate bone at the implantation site. In this sense, chemically grafting organic moieties to the surface of the scaffolds will be further detailed in this chapter. As a consequence of these different approaches, bone tissue engineering is nowadays divided into three general categories: cell-based strategies, growth factor-based strategies, and matrix-based strategies (Mistry and Mikos 2005). However, in real-life situations, two or more of these strategies are frequently combined to implement a solution.

From the material perspective, and in the case of critical size defect scenarios, the implantation of a biomaterial scaffold is essential for filling the defect and stimulating the self-repairing processes of bone. In this sense, the scaffold should be able to deliver biological factors that will promote bone regeneration. Additionally, the scaffold should fulfill profoundly challenging biological and biomechanical functions. In general, ideal scaffolds must fulfill many design requirements (Figure 5.5), such as: providing temporary mechanical support, favoring osteoid deposition, presenting a porous architecture so vascularization and bone in-growth can take place, promoting bone cell migration into the scaffold, encouraging osteoinduction (osteogenic differentiation in the scaffold), stimulating osseointegration-enhancing cellular activity toward scaffold-host tissue integration, controlling the degradation capabilities to favor load transfer to the new developing bone, producing biocompatible degradation products, avoiding any type of inflammatory response, presenting easy sterilizability, and being capable of releasing bioactive molecules and/or drugs to accelerate healing and/or prevent any type of pathology (Porter, Ruckh, and Popat 2009).

5.2.4 BONE REGENERATION PROCESS

Of all the employed materials for the fabrication of 3D scaffolds for the treatment of bone injures, ceramics are excellent candidates since their composition is similar to the inorganic content of living bone, as has been detailed in Chapter 1 of this book. When bioceramics are employed for bone defect regeneration, it is vital that there be an absence of any adverse response from the body as a consequence of the interaction between the implant and the natural tissue, which could result in a fibrous capsule formation around the implant, rendering it ineffective. In a perfect scenario, the bioceramic implant surface should first adsorb the necessary proteins that would promote further cellular adhesion.

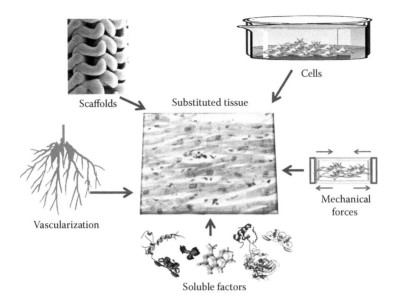

FIGURE 5.5 Schematic representation of the duties that any scaffold should perform in regenerative medicine.

In the next step, the scaffold material should be able to encourage the migration of the mesenchymal bone cells present in the area and promote the osteogenic differentiation to osteoblasts, in a process called osteoinduction (Figure 5.6). Bioceramics scaffolds can promote this osteoinduction process by chemically grafting tissue-inductive substances on their surface or by a controlled release of those substances from their porous structure, as will be detailed later in this chapter.

In the final stage of the bone regeneration process, the osteoblast activity will produce newly formed bone tissue on the surface of the scaffold, in a process called osseointegration. The scaffold should present interconnected macroporosity

Scaffolds fabrication requirements:

➡ Reduction of the immune reaction
➡ Allow cell proliferation
➡ Sensible to cells necessities
➡ Good degradability

FIGURE 5.6 Some of the most important scaffold fabrication requirements.

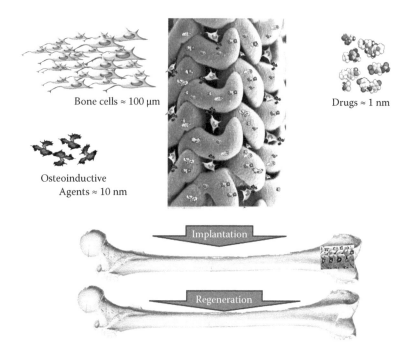

Bone cells ≈ 100 μm

Drugs ≈ 1 nm

Osteoinductive
Agents ≈ 10 nm

Implantation

Regeneration

FIGURE 5.7 **(See color insert.)** Schematic representation of scaffolds with osteoinductive agents on their surface that promote cell attachment and proliferation, and also are able to release certain drugs for diverse bone pathological situations.

to allow vascularization, that is, formation of blood vessels and capillaries, which is one of the most important episodes in bone regeneration process and subsequent bone in-growth. To favor this process, it is possible to add to the scaffold different bioactive agents that promote the adhesion of endothelial progenitor cells (Figure 5.7)

Finally, the scaffold should degrade over time in a controlled manner to allow gradual load transfer to the new bone so the defect will be filled with natural bone. In this sense, and taking into account the whole process, special care should be taken during the scaffold design, particularly in the chemical composition and porous architecture.

5.2.5 THREE-DIMENSIONAL SCAFFOLDS

As has been stated previously, the 3D construct for bone tissue engineering should promote bone tissue growth in the interconnected pore architecture. Two important characteristics can be drawn from that requirement: (1) the porosity should be interconnected to allow vascularization; and (2) the composition should promote natural tissue growing.

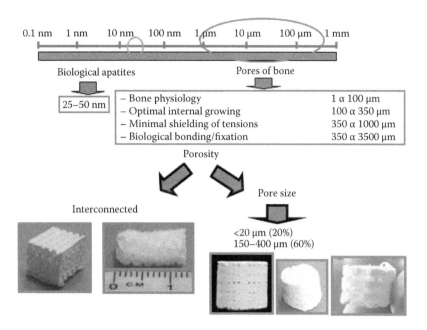

FIGURE 5.8 Required porosity for a 3D scaffold.

5.2.6 POROSITY OF 3D SCAFFOLDS

It is well known that to allow a proper neovascularization and cell infiltration, the pore size and interconnectivity must be larger than 100 μm and there must be a total porosity of 90% (J. X. Lu et al. 1999; Chu et al. 2002; Karageorgiou and Kaplan 2005). Figure 5.8 describes the required porosity of a 3D scaffold.

Porosity in scaffold materials has been traditionally produced by employing porogens that can melt, such as waxes (Oliveira et al. 2009); dissolve, such as sugar and salt (Charles-Harris et al. 2007); or produce carbon dioxide bubbles during the synthetic process (S. S. Kim et al. 2006). Other common methods to producing porous scaffolds include electrospinning, supercritical processing, and freeze-drying from suspensions, using replicas of porous sponges (Fabbri, Celotti, and Ravaglioli 1995), as fused and rapid laser deposition (Juhasz and Best 2012).

However, the best control on scaffold porosity has been obtained using rapid prototyping 3D printing techniques (Leong, Cheah, and Chua 2003). The control and accuracy allow the preparation of 3D scaffolds with complex geometries and fine structures together with porosity control. It is thus possible to design suitable bioceramic scaffolds with the tomography information of the tissue to be replaced for each individual situation. The synthetic process, schematically described in Figure 5.9, starts with the preparation of a paste with the selected ceramic to load a robotic deposition device that extrudes the gel paste layer by layer to manufacture 3D objects according to a previous computer-aided design (Peltola et al. 2008).

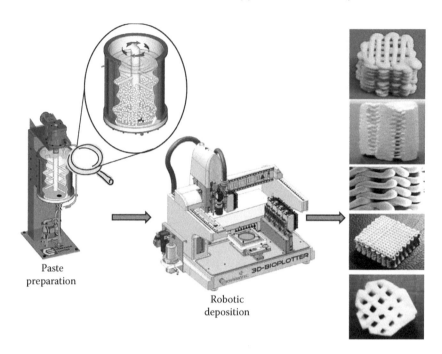

Paste
preparation

Robotic
deposition

FIGURE 5.9 Schematic representation of rapid prototyping 3D printing techniques to produce 3D scaffolds.

Following this synthetic method, it is possible to design and process a porous biodegradable 3D scaffold presenting high porosity, high pore connectivity, and uniform pore distribution. However, to fulfill all the requirements of tissue engineering, it should be possible to release from these scaffolds certain biomolecules, such as growth factors, to stimulate cells to migrate to the local injury site and trigger the healing process. Several attempts have been made by introducing several agents into bioceramic scaffolds to enhance their osteoinductive and vascularization properties (Habraken, Wolke, and Jansen 2007; Mourino and Boccaccini 2010). Among those bioactive agents, there are growth factors like bone morphogenic proteins (BMPs), transforming growth factors (TGFs), basic fibroblast growth factors (bFGFs), and vascular endothelial growth factors (VEGFs). However, those bioactive agents are commonly incorporated through conventional impregnation methods once the scaffolds are produced. The high porosity and interconnected macropores are unable to retain the cargo at the site for a long time, which results in a lack of control over the release kinetics. An interesting approach to solving this problem was the modification of the surface of the scaffold material with certain peptides, such as cell-binding peptides, to enhance cell proliferation and growth. Basically, the incorporation of peptides, such as RGD sequences, onto the surface of the scaffold favors the initial cell reaction and enhances differentiation, proliferation, and mineralization, in the so-called approach of biomimetic materials for tissue engineering (Shin, Jo, and Mikos 2003).

5.2.7 FUNCTIONALIZATION OF 3D SCAFFOLDS

As described previously when dealing with the latest trends in tissue engineering, it is highly recommended that some tissue-inducting substances, such as peptides, hormones, or growth factors, be chemically grafted or physically adsorbed to the scaffold surface to induce the formation of bone tissue. In the case of hydroxyapatite scaffolds, which were produced employing rapid prototyping techniques, the addition of certain peptides to promote bone formation has been positively evaluated. In this sense, the peptide called osteostatin, the 107-111 domain of parathyroid hormone-related protein (PTHrP), has been employed for those studies. The PTHrP protein and the PTH receptor 1 system, which is abundant in both chondrocytes and osteoblasts, are very important in the regulation of bone remodeling (Bisello, Horwitz, and Stewart 2004). It has been observed that the native C-terminal PTHrP (107-139) fragment can inhibit bone resorption acting on osteoclast growth and/or differentiation, and this effect has been ascribed to its N-terminal sequence 107-111 (Thr-Arg-Ser-Ala-Trp), which has been called osteostatin (Rihani Basharat and Lewinson 1997). This peptide was added to Si-doped hydroxyapatite scaffolds to evaluate their osteogenic capabilities. As mentioned earlier, there are two different ways to incorporate bioactive agents to the scaffold: through chemical grafting to previously anchored amine groups or physical adsorption to be then released to their environment (Manzano et al. 2011; Lozano et al. 2011), as shown in Figure 5.10.

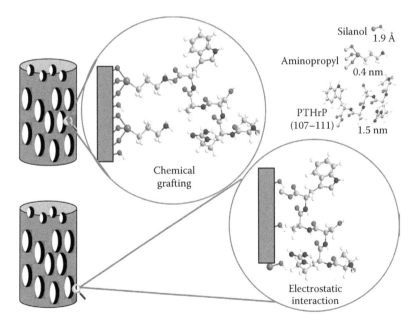

FIGURE 5.10 Chemical grafting of peptides on the surface of silicon-doped hydroxyapatite scaffolds (top); and physical adsorption of peptides into the walls of silicon-doped hydroxyapatite scaffolds (bottom).

The osteogenic action of these scaffolds was evaluated through cell proliferation and viability analyses, together with evaluation of the alkaline phosphatase activity, collagen production, and mineralization in MC3T3-E1 osteoblastic cells. At short periods of time, it was found that osteostatin-adsorbed scaffolds presented an increased cell proliferation. This is in agreement with the peptide release concept, since there is an increment in the local concentration of free osteostatin as a consequence of the release from the scaffold, although there was no control on the kinetics release. Consequently, osteoblasts would be exposed to more peptide and, subsequently, would proliferate faster. However, after longer periods of time, the mineralization was increased in both types of osteostatin-loaded scaffolds.

Another possible solution to the lack of control on the biomolecules' release from tissue-engineering scaffolds could be the use of ordered mesoporous materials to manufacture those 3D scaffolds. As has been detailed throughout this book, ordered mesoporous materials are excellent candidates for drug delivery technologies. Thus using them to construct the macroporous scaffolds would be a real example of using mesoporous materials for tissue engineering. In this sense, the idea is producing scaffolds combining macroporosity, for bone oxygenation and vascularization, and mesoporosity, for biologically active molecule release technology. These scaffolds would be able to guide cell growth and then give rise to the growth of new tissue. However, this is currently only in the research and development step, with mesoporous materials being used as bone regenerative materials rather than tissue-engineering constructs, and it might take a few years to find out the perfect conditions to accomplish it. This is not a trivial challenge, since the adequate conformation method needs to be selected without sacrificing the mechanical properties of the tissue engineering construct. The road map would be the same as that taken by other bioceramics, in which the first step was to investigate the material itself for bone regeneration and then construct the 3D tissue-engineering scaffold with the most successful approach.

The first step of the tissue engineering with silica mesoporous materials road map has already been accomplished with the use of SBA-15 matrices as bone regeneration materials. In this sense, osteostatin, the same peptide as employed with hydroxyapatite scaffolds, has been loaded into SBA-15 materials, and the osteogenic capacity of the matrices has been observed both in vitro (Lozano et al. 2010) with MC3T3-E1 osteoblast cell line and in vivo in a rabbit femur cavity defect model (Trejo et al. 2010) and in a rabbit osteopenia model (Lozano et al. 2012). These preliminary results in which SBA-15 has been shown to promote new bone formation both in vitro and in vivo are very promising for the use of silica mesoporous materials for bone repair and regeneration technologies.

The work that needs to be done with silica mesoporous materials has already been performed with bioactive glasses in the production of 3D bone tissue porous scaffolds based on mesoporous bioactive glasses (Yun, Kim, and Hyeon 2007; Yun et al. 2007, 2008; Li et al. 2007; Zhu et al. 2008; Yun, Kim, and Hyun 2008, 2009; Yun et al. 2008; Li et al. 2008; Yun, Kim, and Park 2011). As mentioned in Chapter 1 of this book, mesoporous bioactive glasses can easily be obtained through a polymer-templating route. Thus it is possible to incorporate different biomolecules into the mesoporous cavities of the scaffold material to be locally released, promoting bone formation or treatment of diverse bone pathologies such as osteoporosis or bone cancer. At the same time, these tissue-engineering constructs present porosity with a wide pore size range and three-dimensional interconnected pore structures to favor blood supply, nutrient delivery, and gas exchange during the in-growth of bone cells. As mentioned previously, the large and interconnected porosity is produced using rapid prototyping procedures. Consequently, every porosity level would play a complementary role in these multifunctional scaffolds.

One of the great advantages of using glasses as the starting materials for the fabrication of those scaffolds is that their chemical composition can be varied depending on the final applications in the clinic. Thus scaffolds can be produced with only SiO_2-P_2O_5 mesoporous materials, which results in a better biocompatible response and less cellular damage than pure silica materials (Garcia et al. 2009). As in the case of pure silica mesoporous materials, the acquired knowledge of these bioactive glasses can be applied to the fabrication of 3D constructs made of SiO_2-P_2O_5 mesoporous glasses and with three ranges of porosity: (1) meso-sized pores, with pores of 4 nm resulting from polymer templating during the synthetic process of the glass; (2) macro-sized pores, with pores in the range of 30–80 μm as a consequence of using methylcellulose as porogens; and (3) giant-sized pores, with pores of ca. 400 μm produced using rapid prototyping technologies (Garcia et al. 2011). Thus these tissue engineering constructs follow most of the requirements for a scaffold to be used in tissue engineering technologies, since they can promote cell seeding, migration, and tissue in-growth.

However, these hierarchically giant-, macro-, and mesoporous 3D scaffolds made of bioactive glasses for tissue engineering normally present poor mechanical properties because they are too brittle. In this sense, to improve those mechanical properties it is possible to add certain polymers to the composition, such as poly(caprolactone) to obtain composite scaffolds with sponge-like mechanical properties (Yun, Kim, and Park 2011). These polymer-glass composite scaffolds are of great interest for use in tissue engineering technologies because of their shapability, bioactive behavior, and adjustable degradation rate (Rezwan et al. 2006).

5.3 BIOCOMPATIBILITY OF ORDERED MESOPOROUS SILICA

During the evaluation of the biocompatibility of any biomaterial, a distinction should be considered depending on the function to be performed by the material itself. The physiological environment where the biomaterial will perform its role will be different depending on the administration route to the living body. The two main biomedical applications of silica mesoporous materials should be highlighted: as implantable devices for bone tissue engineering and as nanoparticles for drug delivery technologies.

5.3.1 BIOCOMPATIBILITY OF ORDERED MESOPOROUS SILICA AS IMPLANTABLE DEVICES

As has been detailed in Chapter 1 of this book, ordered mesoporous silica materials present a similar chemical surface to bioglasses, which have been used as bone regenerators. This fact was the inspiration for using these mesoporous silica materials as implantable materials for bone regeneration. They were observed to develop nanoapatite coatings with a crystallinity similar to that of biological apatites when immersed in physiological fluids (Vallet-Regí et al. 2006). This pioneering work suggested these materials for bone regeneration technologies, since the implanted material would bond to living bone in the body.

5.3.1.1   In Vitro Studies

As has been already mentioned in this chapter, ordered mesoporous materials can be loaded with different peptides to promote bone formation. That was the case with SBA-15 materials, unmodified and organically modified, loaded with osteostatin, the 107-111 domain of parathyroid hormone-related protein (Lozano et al. 2010). These materials were evaluated in vitro using osteoblastic cells, and it was observed that the release of the peptide from the matrix promoted cell growth and differentiation, and also the expression of different osteoblastic products, such as alkaline phosphatase, osteocalcin, collagen, osteoprotegrin, receptor activator of nuclear factor-κB ligand, and vascular endothelial growth factor. The obtained results encouraged us to carry on the research with further in vivo studies, as detailed in the next subsection.

5.3.2.1   In Vivo Studies

When the same materials mentioned in the previous subsection, that is, osteostatin-loaded SBA-15 matrices, were implanted in different animal models, very promising results were obtained. They were implanted in a cavitary defect in the rabbit femur (Trejo et al. 2010) and in a rabbit osteopenia model (Lozano et al. 2012), obtaining very positive results in both animal models. Those experiments showed that the addition of osteostatin to SBA-15 materials improved local bone induction, promoting an increase in the early repair response, not only in normal bone but also in osteoporotic bone after local injury.

Thus the next step in this research should be the design of 3D scaffolds with ordered mesoporous material, so a double function can be performed (Figure 5.11): bone tissue regeneration promoted by peptides and growth factors anchored and/or physically adsorbed to the walls of the scaffold itself; and drug delivery for the treatment of diverse pathologies thanks to the mesoporous porosity and drug delivery abilities from ordered mesoporous materials.

5.3.2   BIOCOMPATIBILITY OF ORDERED MESOPOROUS SILICA AS NANOPARTICLES FOR DRUG DELIVERY TECHNOLOGIES

When considering mesoporous silica nanoparticles for biomedicine applications, safety and toxicological issues are of great importance. Attending to the chemical composition, ordered mesoporous silica presents the same chemical composition as synthetic amorphous silica, which is an approved food additive by EU regulations (Garcia-Bennett 2011). Moreover, amorphous silica is commonly recognized as nontoxic, and it has been traditionally considered as biocompatible and degradable in living tissues (Martin 2007). Additionally, amorphous colloidal and porous silica are

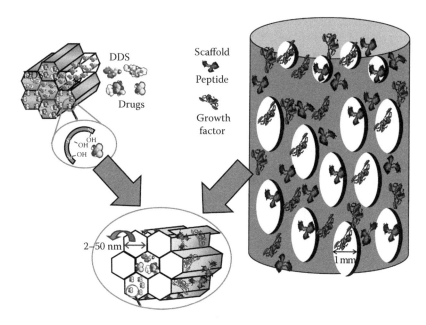

FIGURE 5.11 **(See color insert.)** Future scaffolds made of ordered mesoporous materials with two functions: bone tissue regeneration and drug delivery capabilities.

used as adjuvants in pharmaceutical technology. However, the textural properties of ordered mesoporous silicas, with high surface area and large porosity, make these materials different from amorphous silica. In fact, their high surface area promotes adsorbing a large quantity of biomolecules on their surface, which is the first step when interacting with living tissues. Thus this different behavior makes obligatory the evaluation of their immunotoxicological characteristics without taking into consideration amorphous silica. Additionally, it is well known in the nanotechnology area that the toxic effects of nanoparticles depend on the physicochemical properties of the employed materials, including size, charge, and coating ligands (Schroeder et al. 2012).

5.3.2.1 In Vitro Studies

It has been already established that within biomaterials technology, and particularly regarding toxicity matters, there is not a direct correlation between in vitro and in vivo situations (Kohane and Langer 2010), which make in vivo studies essential for understanding the biocompatibility of a given material. However, in vitro studies, which provide basic information of the ability of certain cell types to survive in the presence of a biomaterial, are an indispensable tool for discarding possible toxic materials. Moreover, the number of publications regarding the possible toxicity of any type of nanoparticle has demonstrated that an evaluation of the in vitro cytotoxicity of nanoparticles is essential before their medical application.

The in vitro biocompatibility of ordered mesoporous nanoparticles has been positively evaluated using many different cell lines, as is detailed in the following.

Interestingly, it has been found that the concentration of those nanoparticles is a key factor in their in vitro toxicity; those studies showed a concentration-dependent phenomenon, with low toxicity at low concentrations and high toxicity at high concentrations (Vivero-Escoto et al. 2010). Additionally, it has been found that the possible residual fragments of degraded surfactants coming from the thermal removal can be cytotoxic. In this sense, in silica mesoporous nanoparticles for biomedical uses the surfactant is normally removed by solvent extraction rather than by calcination (Manzano, Colilla, and Vallet-Regí 2009).

It has been observed that by controlling certain physicochemical properties of mesoporous silica nanoparticles, such as particle size, shape, and surface chemistry, they can be efficiently endocytosed by living cells with high in vitro biocompatibility, and they can escape the endolysosomal entrapment (Lin et al. 2005). Those results make these nanoparticles excellent vectors for transporting any type of biomolecules into living cells and releasing them into their cytoplasm. In this sense, ordered mesoporous nanoparticles have been efficiently endocytosed in vitro by a variety of mammalian cells, such as cancer cells (human cervical cancer cells) (Slowing, Trewyn, and Lin 2006, 2007; Radu et al. 2004; Giri et al. 2005; F. Lu et al. 2009), pancreatic cancer cells (J. Lu et al. 2007), Chinese hamster ovarian cells (Radu et al. 2004), non-small-cell lung cancer epithelial cells (Yu, Malugin, and Ghandehari 2011), noncancer (neurological) cells (Radu et al. 2004; Lai et al. 2003), macrophages (primarily human monocyte-derived macrophages) (Yu, Malugin, and Ghandehari 2011; Witasp et al. 2009), human mesenchymal stem cells (Chung et al. 2007; Huang et al. 2005), human dendritic cells (Vallhov et al. 2007), and others (Tao et al. 2008). An important finding considering the potential application of mesoporous silica nanoparticles as nanocarriers through the bloodstream was that these nanovectors are nontoxic toward mammalian red blood cells and, what is more important, present a good hemocompatibility (Slowing et al. 2009; Lin and Haynes 2010).

The increasing research on nanotechnology for drug delivery has found that particle shape and size are very important factors for a carrier to be efficiently uptaken by nonphagocytic cells (Slowing et al. 2008). As has been detailed in the previous chapters of this book, mesoporous silica nanoparticles bear a straightforward advantage over other types of nanocarriers since they can be fabricated in the submicron scale with precise control of the particle size. Moreover, it is also well known that surface properties of the nanoparticles are also an important point, and as described throughout this book, mesoporous silica nanoparticles are very robust and can be easily modified on their internal and external surface with the required organic moiety. Consequently, mesoporous silica nanoparticles with adequate size and surface functionalization, can offer an increased circulation time in the bloodstream when intravenously injected, since they can be engineered to be stealth agents in the immune system.

5.2.3.2 In Vivo Studies: Biodistribution, Retention, and Clearance

Although the in vitro experiments reveal only the behavior of certain cells under the presence of a given material, it is obvious that all the factors affecting the in vitro toxicity and biocompatibility, such as particle size and shape together with surface area and functionalization, will influence their in vivo response (Lopez et al. 2006; Wu et al. 2008). However, the research on in vivo behavior of mesoporous silica nanoparticles has just started in the last few years. As a consequence, there are controversies and debates about the toxicological results (Garcia-Bennett 2011). A recent investigation on mesoporous silica nanoparticle biocompatibility and biodistribution has revealed that when the size range of those nanoparticles was between 100 and 130 nm, they were well tolerated when injected intravenously, as serological, hematological, and histopathological examinations of blood samples and mouse tissues showed (Lu et al. 2010). Similarly, and as detailed in the previous chapters of this book, mesoporous silica nanoparticles have been explored as in vivo bioimaging agents without any toxicity in the short term (Liu et al. 2008; Hsiao et al. 2008; J. Kim et al. 2008; Taylor et al. 2008).

An important factor in drug delivery technologies is the route of administration of the designed nanoparticles, especially regarding the potential toxicity. In this sense, mesoporous silica nanoparticles have shown a good biocompatibility in rats when administered subcutaneously (Hudson et al. 2008). On the other hand, using a different administration route, such as intraperitoneal and intravenous injections, on mice resulted in death or euthanasia. This effect can be mitigated by a proper surface functionalization of the nanoparticles or using lower concentrations during the in vivo experiments. This controversy denotes that there is a considerable need for the completion of more toxicological research before these nanoparticles could be developed in clinical phases. However, their great versatility would allow the production of nanoparticles with better blood circulation, biodegradability, and clearance from the body.

REFERENCES

Bisello, A., M. J. Horwitz, and A. F. Stewart. 2004. "Parathyroid Hormone-Related Protein: An Essential Physiological Regulator of Adult Bone Mass." *Endocrinology* 145 (8):3551–53.

Charles-Harris, M., S. del Valle, E. Hentges, P. Bleuet, D. Lacroix, and J. A. Planell. 2007. "Mechanical and Structural Characterisation of Completely Degradable Polylactic Acid/Calcium Phosphate Glass Scaffolds." *Biomaterials* 28 (30):4429–38.

Chu, T. M. G., D. G. Orton, S. J. Hollister, S. E. Feinberg, and J. W. Halloran. 2002. "Mechanical and In Vivo Performance of Hydroxyapatite Implants with Controlled Architectures." *Biomaterials* 23 (5):1283–93.

Chung, T. H., S. H. Wu, M. Yao, C. W. Lu, Y. S. Lin, Y. Hung, C. Y. Mou, Y. C. Chen, and D. M. Huang. 2007. "The Effect of Surface Charge on the Uptake and Biological Function of Mesoporous Silica Nanoparticles 3T3-L1 Cells and Human Mesenchymal Stem Cells." *Biomaterials* 28 (19):2959–66.

Fabbri, M., G. C. Celotti, and A. Ravaglioli. 1995. "Hydroxyapatite-Based Porous Aggregates: Physico-Chemical Nature, Structure, Texture and Architecture." *Biomaterials* 16 (3):225–28.

Garcia, A., M. Colilla, I. Izquierdo-Barba, and M. Vallet-Regí. 2009. "Incorporation of Phosphorus into Mesostructured Silicas: A Novel Approach to Reduce the SiO(2) Leaching in Water." *Chemistry of Materials* 21 (18):4135–45.

Garcia, A., I. Izquierdo-Barba, M. Colilla, C. L. de Laorden, and M. Vallet-Regí. 2011. "Preparation of 3-D Scaffolds in the SiO(2)-P(2)O(5) System with Tailored Hierarchical Meso-Macroporosity." *Acta Biomaterialia* 7 (3):1265–73.

Garcia-Bennett, A. E. 2011. "Synthesis, Toxicology and Potential of Ordered Mesoporous Materials in Nanomedicine." *Nanomedicine* 6 (5):867–77.

Giri, S., B. G. Trewyn, M. P. Stellmaker, and V. S. Y. Lin. 2005. "Stimuli-Responsive Controlled-Release Delivery System Based on Mesoporous Silica Nanorods Capped with Magnetic Nanoparticles." *Angewandte Chemie-International Edition* 44 (32):5038–44.

Habraken, Wjem, J. G. C. Wolke, and J. A. Jansen. 2007. "Ceramic Composites as Matrices and Scaffolds for Drug Delivery in Tissue Engineering." *Advanced Drug Delivery Reviews* 59 (4–5):234–48.

Hsiao, J. K., C. P. Tsai, T. H. Chung, Y. Hung, M. Yao, H. M. Liu, C. Y. Mou, C. S. Yang, Y. C. Chen, and D. M. Huang. 2008. "Mesoporous Silica Nanoparticles as a Delivery System of Gadolinium for Effective Human Stem Cell Tracking." *Small* 4 (9):1445–52.

Huang, Dong-Ming, Yann Hung, Bor-Sheng Ko, Szu-Chun Hsu, Wei-Hsuan Chen, Chung-Liang Chien, Chih-Pin Tsai, Chieh-Ti Kuo, Ju-Chiun Kang, Chung-Shi Yang, Chung-Yuan Mou, and Yao-Chang Chen. 2005. "Highly Efficient Cellular Labeling of Mesoporous Nanoparticles in Human Mesenchymal Stem Cells: Implication for Stem Cell Tracking." *The FASEB Journal* 19 (14):2014–16.

Hudson, S. P., R. F. Padera, R. Langer, and D. S. Kohane. 2008. "The Biocompatibility of Mesoporous Silicates." *Biomaterials* 29 (30):4045–55.

Juhasz, J. A., and S. M. Best. 2012. "Bioactive Ceramics: Processing, Structures and Properties." *Journal of Materials Science* 47 (2):610–24.

Karageorgiou, V., and D. Kaplan. 2005. "Porosity of 3D Biormaterial Scaffolds and Osteogenesis." *Biomaterials* 26 (27):5474–91.

Kim, Jaeyun, Hoe Suk Kim, Nohyun Lee, Taeho Kim, Hyoungsu Kim, Taekyung Yu, In Chan Song, Woo Kyung Moon, and Taeghwan Hyeon. 2008. "Multifunctional Uniform Nanoparticles Composed of a Magnetite Nanocrystal Core and a Mesoporous Silica Shell for Magnetic Resonance and Fluorescence Imaging and for Drug Delivery." *Angewandte Chemie International Edition* 47 (44):8438–41.

Kim, S. S., M. S. Park, O. Jeon, C. Y. Choi, and B. S. Kim. 2006. "Poly(lactide-co-glycolide)/Hydroxyapatite Composite Scaffolds for Bone Tissue Engineering." *Biomaterials* 27.

Kohane, D. S., and R. Langer. 2010. "Biocompatibility and Drug Delivery Systems." *Chemical Science* 1 (4):441–46.

Lai, C. Y., B. G. Trewyn, D. M. Jeftinija, K. Jeftinija, S. Xu, S. Jeftinija, and V. S. Y. Lin. 2003. "A Mesoporous Silica Nanosphere-Based Carrier System with Chemically Removable CdS Nanoparticle Caps for Stimuli-Responsive Controlled Release of Neurotransmitters and Drug Molecules." *Journal of the American Chemical Society* 125 (15):4451–59.

Langer, R. 2000. "Biomaterials in Drug Delivery and Tissue Engineering: One Laboratory's Experience." *Accounts of Chemical Research* 33 (2):94–101.

Langer, R., and J. P. Vacanti. 1993. "Tissue Engineering." *Science* 260 (5110):920–26.

Leong, K. F., C. M. Cheah, and C. K. Chua. 2003. "Solid Freeform Fabrication of Three-Dimensional Scaffolds for Engineering Replacement Tissues and Organs." *Biomaterials* 24 (13):2363–78.

Li, X., X. P. Wang, H. R. Chen, P. Jiang, X. P. Dong, and J. L. Shi. 2007. "Hierarchically Porous Bioactive Glass Scaffolds Synthesized with a PUF and P123 Cotemplated Approach." *Chemistry of Materials* 19 (17):4322–26.

Li, Xia, Jianlin Shi, Xiaoping Dong, Lingxia Zhang, and Hongyu Zeng. 2008. "A Mesoporous Bioactive Glass/Polycaprolactone Composite Scaffold and Its Bioactivity Behavior." *Journal of Biomedical Materials Research Part A* 84A (1):84–91.

Lin, Y. S., and C. L. Haynes. 2010. "Impacts of Mesoporous Silica Nanoparticle Size, Pore Ordering, and Pore Integrity on Hemolytic Activity." *Journal of the American Chemical Society* 132 (13):4834–42.

Lin, Y. S., C. P. Tsai, H. Y. Huang, C. T. Kuo, Y. Hung, D. M. Huang, Y. C. Chen, and C. Y. Mou. 2005. "Well-Ordered Mesoporous Silica Nanoparticles as Cell Markers." *Chemistry of Materials* 17 (18):4570–73.

Liu, H. M., S. H. Wu, C. W. Lu, M. Yao, J. K. Hsiao, Y. Hung, Y. S. Lin, C. Y. Mou, C. S. Yang, D. M. Huang, and Y. C. Chen. 2008. "Mesoporous Silica Nanoparticles Improve Magnetic Labeling Efficiency in Human Stem Cells." *Small* 4 (5):619–26.

Lopez, T., E. I. Basaldella, M. L. Ojeda, J. Manjarrez, and R. Alexander-Katz. 2006. "Encapsulation of Valproic Acid and Sodic Phenytoin in Ordered Mesoporous SiO2 Solids for the Treatment of Temporal Lobe Epilepsy." *Optical Materials* 29 (1):75–81.

Lozano, D., M. Manzano, D. Arcos, S. Portal-Nunez, P. Esbrit, and M. Vallet-Regí. 2011. "Si-hydroxyapatite with Covalently Linked or Adsorbed Osteostatin Exhibits Improved Osteogenic Capacity in Osteoblastic Cells." *Osteoporosis International* 22:329–30.

Lozano, D., M. Manzano, J. C. Doadrio, A. J. Salinas, M. Vallet-Regí, E. Gomez-Barrena, and P. Esbrit. 2010. "Osteostatin-Loaded Bioceramics Stimulate Osteoblastic Growth and Differentiation." *Acta Biomaterialia* 6 (3):797–803.

Lozano, Daniel, Cynthia G. Trejo, Enrique Gómez-Barrena, Miguel Manzano, Juan C. Doadrio, Antonio J. Salinas, María Vallet-Regí, Natalio García-Honduvilla, Pedro Esbrit, and Julia Buján. 2012. "Osteostatin-Loaded onto Mesoporous Ceramics Improves the Early Phase of Bone Regeneration in a Rabbit Osteopenia Model." *Acta Biomaterialia* 8 (6):2317–23.

Lu, F., S. H. Wu, Y. Hung, and C. Y. Mou. 2009. "Size Effect on Cell Uptake in Well-Suspended, Uniform Mesoporous Silica Nanoparticles." *Small* 5 (12):1408–13.

Lu, J., M. Liong, Z. X. Li, J. I. Zink, and F. Tamanoi. 2010. "Biocompatibility, Biodistribution, and Drug-Delivery Efficiency of Mesoporous Silica Nanoparticles for Cancer Therapy in Animals." *Small* 6 (16):1794–1805.

Lu, J., M. Liong, J. I. Zink, and F. Tamanoi. 2007. "Mesoporous Silica Nanoparticles as a Delivery System for Hydrophobic Anticancer Drugs." *Small* 3 (8):1341–46.

Lu, J. X., B. Flautre, K. Anselme, P. Hardouin, A. Gallur, M. Descamps, and B. Thierry. 1999. "Role of Interconnections in Porous Bioceramics on Bone Recolonization In Vitro and In Vivo." *Journal of Materials Science-Materials in Medicine* 10 (2):111–20.

Manzano, M., M. Colilla, and M. Vallet-Regí. 2009. "Drug Delivery from Ordered Mesoporous Matrices." *Expert Opinion on Drug Delivery* 6 (12):1383–1400.

Manzano, M., D. Lozano, D. Arcos, S. Portal-Nunez, C. L. la Orden, P. Esbrit, and M. Vallet-Regí. 2011. "Comparison of the Osteoblastic Activity Conferred on Si-Doped Hydroxyapatite Scaffolds by Different Osteostatin Coatings." *Acta Biomaterialia* 7 (10):3555–62.

Martin, K. R. 2007. "The Chemistry of Silica and Its Potential Health Benefits." *Journal of Nutrition Health & Aging* 11 (2):94–98.

Mistry, A. S., and A. G. Mikos. 2005. "Tissue Engineering Strategies for Bone Regeneration." In *Regenerative Medicine II: Clinical and Preclinical Applications*, edited I. V. Yannas, 1–22. Berlin: Springer.

Mouriño, V., and A. R. Boccaccini. 2010. "Bone Tissue Engineering Therapeutics: Controlled Drug Delivery in Three-Dimensional Scaffolds." *Journal of the Royal Society Interface* 7 (43):209–27.

O'Brien, Fergal J. 2011. "Biomaterials & Scaffolds for Tissue Engineering." *Materials Today* 14 (3):88–95.

Oliveira, J. M., S. A. Costa, I. B. Leonor, P. B. Malafaya, J. F. Mano, and R. L. Reis. 2009. "Novel Hydroxyapatite/Carboxymethylchitosan Composite Scaffolds Prepared through an Innovative 'Autocatalytic' Electroless Coprecipitation Route." *Journal of Biomedical Materials Research Part A* 88A (2):470–80.

Peltola, S. M., F. P. W. Melchels, D. W. Grijpma, and M. Kellomaki. 2008. "A Review of Rapid Prototyping Techniques for Tissue Engineering Purposes." *Annals of Medicine* 40 (4):268–80.

Place, E. S., N. D. Evans, and M. M. Stevens. 2009. "Complexity in Biomaterials for Tissue Engineering." *Nature Materials* 8 (6):457–70.

Porter, Joshua R., Timothy T. Ruckh, and Ketul C. Popat. 2009. "Bone Tissue Engineering: A Review in Bone Biomimetics and Drug Delivery Strategies." *Biotechnology Progress* 25 (6):1539–60.

Radu, D. R., C. Y. Lai, K. Jeftinija, E. W. Rowe, S. Jeftinija, and V. S. Y. Lin. 2004. "A Polyamidoamine Dendrimer-Capped Mesoporous Silica Nanosphere-Based Gene Transfection Reagent." *Journal of the American Chemical Society* 126 (41):13216–17.

Rezwan, K., Q. Z. Chen, J. J. Blaker, and A. R. Boccaccini. 2006. "Biodegradable and Bioactive Porous Polymer/Inorganic Composite Scaffolds for Bone Tissue Engineering." *Biomaterials* 27 (18):3413–31.

Rihani Basharat, S., and D. Lewinson. 1997. "PTHrP(107-111) Inhibits In Vivo Resorption That Was Stimulated by PTHrP(1-34) When Applied Intermittently to Neonatal Mice." *Calcified Tissue International* 61 (5):426–28.

Schroeder, Avi, Daniel A. Heller, Monte M. Winslow, James E. Dahlman, George W. Pratt, Robert Langer, Tyler Jacks, and Daniel G. Anderson. 2012. "Treating Metastatic Cancer with Nanotechnology." *Nature Reviews Cancer* 12 (1):39–50.

Shin, H., S. Jo, and A. G. Mikos. 2003. "Biomimetic Materials for Tissue Engineering." *Biomaterials* 24 (24):4353–64.

Slowing, I., B. G. Trewyn, and V. S. Y. Lin. 2006. "Effect of Surface Functionalization of MCM-41-Type Mesoporous Silica Nanoparticles on the Endocytosis by Human Cancer Cells." *Journal of the American Chemical Society* 128 (46):14792–93.

Slowing, I. I., B. G. Trewyn, and V. S. Y. Lin. 2007. "Mesoporous Silica Nanoparticles for Intracellular Delivery of Membrane-Impermeable Proteins." *Journal of the American Chemical Society* 129 (28):8845–49.

Slowing, I. I., J. L. Vivero-Escoto, C. W. Wu, and V. S. Y. Lin. 2008. "Mesoporous Silica Nanoparticles as Controlled Release Drug Delivery and Gene Transfection Carriers." *Advanced Drug Delivery Reviews* 60 (11):1278–88.

Slowing, I. I., C. W. Wu, J. L. Vivero-Escoto, and V. S. Y. Lin. 2009. "Mesoporous Silica Nanoparticles for Reducing Hemolytic Activity Towards Mammalian Red Blood Cells." *Small* 5 (1):57–62.

Tao, Z. M., M. P. Morrow, T. Asefa, K. K. Sharma, C. Duncan, A. Anan, H. S. Penefsky, J. Goodisman, and A. K. Souid. 2008. "Mesoporous Silica Nanoparticles Inhibit Cellular Respiration." *Nano Letters* 8 (5):1517–26.

Taylor, K. M. L., J. S. Kim, W. J. Rieter, H. An, W. L. Lin, and W. B. Lin. 2008. "Mesoporous Silica Nanospheres as Highly Efficient MRI Contrast Agents." *Journal of the American Chemical Society* 130 (7):2154–55.

Trejo, C. G., D. Lozano, M. Manzano, J. C. Doadrio, A. J. Salinas, S. Dapia, E. Gomez-Barrena, M. Vallet-Regi, N. Garcia-Honduvilla, J. Bujan, and P. Esbrit. 2010. "The Osteoinductive Properties of Mesoporous Silicate Coated with Osteostatin in a Rabbit Femur Cavity Defect Model." *Biomaterials* 31 (33):8564–73.

Vallet-Regí, M., and D. Arcos. 2008. "Biological Apatites in Bone and Teeth." In *Biomimetic Nanoceramics in Clinical Use: From Materials to Applications*, 1–24. Cambridge: Royal Society of Chemistry.

Vallet-Regí, M. A., L. Ruiz-González, I. Izquierdo-Barba, and J. M. González-Calbet. 2006. "Revisiting Silica Based Ordered Mesoporous Materials: Medical Applications." *Journal of Materials Chemistry* 16 (1):26–31.

Vallet-Regí, María, and Eduardo Ruiz-Hernández. 2011. "Bioceramics: From Bone Regeneration to Cancer Nanomedicine." *Advanced Materials* 23 (44):5177–5218.

Vallhov, H., S. Gabrielsson, M. Stromme, A. Scheynius, and A. E. Garcia-Bennett. 2007. "Mesoporous Silica Particles Induce Size Dependent Effects on Human Dendritic Cells." *Nano Letters* 7 (12):3576–82.

Vivero-Escoto, J. L., Slowing, II, B. G. Trewyn, and V. S. Y. Lin. 2010. "Mesoporous Silica Nanoparticles for Intracellular Controlled Drug Delivery." *Small* 6 (18):1952–67.

Witasp, E., N. Kupferschmidt, L. Bengtsson, K. Hultenby, C. Smedman, S. Paulie, A. E. Garcia-Bennett, and B. Fadeel. 2009. "Efficient Internalization of Mesoporous Silica Particles of Different Sizes by Primary Human Macrophages without Impairment of Macrophage Clearance of Apoptotic or Antibody-Opsonized Target Cells." *Toxicology and Applied Pharmacology* 239 (3):306–19.

Wu, Si-Han, Yu-Shen Lin, Yann Hung, Yi-Hsin Chou, Yi-Hua Hsu, Chen Chang, and Chung-Yuan Mou. 2008. "Multifunctional Mesoporous Silica Nanoparticles for Intracellular Labeling and Animal Magnetic Resonance Imaging Studies." *ChemBioChem* 9 (1):53–57.

Yu, T., A. Malugin, and H. Ghandehari. 2011. "Impact of Silica Nanoparticle Design on Cellular Toxicity and Hemolytic Activity." *ACS Nano* 5 (7):5717–28.

Yun, H. S., S. E. Kim, and Y. T. Hyeon. 2007. "Design and Preparation of Bioactive Glasses with Hierarchical Pore Networks." *Chemical Communications* (21):2139–41.

Yun, H. S., S. E. Kim, and Y. T. Hyun. 2008. "Fabrication of Hierarchically Porous Bioactive Glass Ceramics." In *Bioceramics* (Vol. 20, Pts. 1 and 2), edited by G. Daculsi and P. Layrolle, 285–288. Stafa-Zurich: Trans Tech Publications.

Yun, H. S., S. E. Kim, and Y. T. Hyun. 2009. "Preparation of Bioactive Glass Ceramic Beads with Hierarchical Pore Structure Using Polymer Self-Assembly Technique." *Materials Chemistry and Physics* 115 (2–3):670–76.

Yun, H. S., S. E. Kim, Y. T. Hyun, S. J. Heo, and J. W. Shin. 2007. "Three-Dimensional Mesoporous-Giantporous Inorganic/Organic Composite Scaffolds for Tissue Engineering." *Chemistry of Materials* 19 (26):6363–66.

Yun, H. S., S. E. Kim, Y. T. Hyun, S. J. Heo, and J. W. Shin. 2008. "Hierarchically Mesoporous-Macroporous Bioactive Glasses Scaffolds for Bone Tissue Regeneration." *Journal of Biomedical Materials Research Part B: Applied Biomaterials* 87B (2):374–80.

Yun, H. S., S. E. Kim, and E. K. Park. 2011. "Bioactive Glass-Poly (Epsilon-Caprolactone) Composite Scaffolds with 3 Dimensionally Hierarchical Pore Networks." *Materials Science & Engineering C—Materials for Biological Applications* 31 (2):198–205.

Zhu, Y. F., C. T. Wu, Y. Ramaswamy, E. Kockrick, P. Simon, S. Kaskel, and H. Zreiqat. 2008. "Preparation, Characterization and In Vitro Bioactivity of Mesoporous Bioactive Glasses (MBGs) Scaffolds for Bone Tissue Engineering." *Microporous and Mesoporous Materials* 112 (1–3):494–503.

Index

Page references in **bold** refer to tables.

A

Active targeting, 142–144
Adsorption
  amino acids, 85
  and carrier properties, 70–72
  characterization techniques, 71–72
  and drug molecule size, 73
  electrostatic, 139
  enzymes, 74
  factors in, 70
  Fourier transform infrared (FTIR)
      spectroscopy, 80
  and functionalization, 81–86
  hydrogen bonding, 138–140
  hydroxyapatite scaffolds, 188, 193
  and matrix surface area, 77–78
  mesoporous silica nanoparticles (MSNPs),
      138–140
  and pH, 139
  and pore size, 72–77
  and pore volume, 70, 78–81
  proteins, 74–75
  steps in, 70–72
  and surface area, 70, 77–78
Adsorption isotherm, 74
Aerosol-assisted synthesis, 136
Aging, sol-gel process, 16
Aggregation, micelles, 4
Albumin, 141
Alcohol condensation, 16
Alendronate, 77–78, 83–84, 88, 89–90
Alkoxysilanes, 81–82
Allografts, 183
Alternating magnetic fields (AMFs), 156–157
Alumina, 1
Amidase, 122
Amine dendrimers, 90–91
Amino acids, 85
3-Aminopropyltrimethoxysilane (APTS), 116
Amorphous calcium phosphate (ACP), 24,
      48–55
Amorphous silica, 95, 196–197
Amorphous solids, 7
α-Amylase, 121
AND logic gates, 127
Anesthetics, topical, 98
Antibiotics, 98
Antibody-antigen interactions, 123

Antigens, as release triggers, 123
Apatite layer formation, 6, 14, 46–47
Apatites, 9
Apatite-wollastonite (A-W) glass-ceramic, 42
Aptamers, 123–126
ATP aptamer, 123–124
Autografts, 183
Avidin caps, 121–122
Azobenzene, 115–116, 127

B

Bcl-2 protein, 148
BDDT (Brunauer-Deming-Deming-Teller)
      method, 74
BET (Brunauer-Emmet-Teller) method, 77
Bioactive ceramics. *See* Bioceramics
Bioactive glasses; *See also* Mesoporous bioactive
      glasses (MBGs)
  applications, 1, **7**, 13
  bioactivity mechanism, xi, 6–7, 13, 14, 40, 42,
      51–55
  characteristics, 13
  coatings, 18–19
  compositions, 14
  crystallization, 7
  kinetics, 13
  melt glasses, 14
  microspheres, 26
  sol-gel, 14–27
Bioactivity
  accelerators, 13
  bioactive glasses, xi, 6–7, 13, 14, 40, 42,
      51–55
  bioceramics, xi, 1, 6–7, 51–55
  calcium and, 17–18, 28, 46–47, 48–49
  defined, xi, 1
  glass-ceramics, xi, 1, 6–7, 21, 40, 42, 51–55
  inducers, 13
  phosphorus and, 17–18, 24, 54
  stages of, 51–55
  surface response, 39, 40, 42–44
Bioactivity assays
  buffered aqueous solution, 39, **41**, 42
  and calcium content, 46–47, 48–49
  carbonated simulated inorganic plasma
      (CSIP), 43
  characterization techniques, 44–45
  dynamic, 43–44
  Hanks' balanced salt solution (HBSS), 43
  ordered mesoporous glasses, 45–55

T - #0417 - 071024 - C11 - 234/156/11 - PB - 9780367380601 - Gloss Lamination